SEMICENTENNIAL PUBLICATIONS

OF THE

UNIVERSITY OF CALIFORNIA

1868-1918

THE FUNDAMENTAL EQUATIONS OF DYNAMICS AND ITS MAIN COÖRDINATE SYSTEMS VECTORIALLY TREATED AND ILLUSTRATED FROM RIGID DYNAMICS

BY

FREDERICK SLATE

UNIVERSITY OF CALIFORNIA PRESS
BERKELEY
1918

THIS BOOK FORMS PART II OF
THE PRINCIPLES OF MECHANICS, PART I,
NEW YORK, THE MACMILLAN COMPANY, 1900

1918

PREFACE

The day has clearly passed when any comprehensive presentation of all dynamics could be compressed and unified within the compass of one moderate volume of homogeneous plan. The established connections of dynamical reasoning with other fields in physics are of increasing number and closeness, as furnishing for them strongly rooted sequences in their interpretative trains of thought and linking together what would else have continued to stand separate. And that relation has reacted powerfully in modern times upon dynamics itself, perpetually enriching its substance, yet at the same time introducing within it certain sharpening differences that are stamped upon it by the type of use for which preparation is being made. These in fact modify superficially the modes of expression and their tone, and shift their own emphasis through a range that brings about what is in effect a subdivision of territory and an acknowledgment of practically diverse interests. It is in response to the situation which has been thus unfolding, and in conformity with its drift toward manifold adaptations, that special treatises have been rendered available whose measure of unquestioned excellence and authority would make superfluous an attempt to replace any such unit with a marked improvement upon it.

But undoubtedly these differentiations founded in divergencies and inevitably expressing them in some degree, are entailing a corresponding need and demand to offset them with a broadening survey of the common foundation and of the common stock of resources. And with that end in view another treatment of dynamics finds a place for itself and holds it for special service. This will propose to state with catholic inclusiveness the principles that lay out and direct all the main lines of use, and to anticipate at their common source, as it were, the preferred methods and forms that are characteristic of various provinces.

On this side also reasonable requirements for the immediate future have been satisfied up to a definitely recognizable point. For works on abstract dynamics are at hand to help, whose number and quality have left no fair opening for renewed exposition, that could indeed scarcely attain excellence without duplicating them. In the same proportion, however, that their requisite perspective has grown, until it involves truly panoramic sweep, its due scope must cease to be secured except from a distance that expunges most details and spares only landmarks of the bolder outlines. And under the urgent pressure to condense in order to avoid neglecting and yet not become too voluminous in summarizing completely, to keep even pace with widening outlook, this view of dynamics cannot but endure the attendant risks of abstractness. Because it must lean in building toward great reliance upon the formal aid of mathematics, perforce the physical coloring will fade and the bonds with experimental reasoning be loosened. The stated results are progressively less likely to comprise what is charged with tentative quality and is held with the candidly provisional acceptance proper to inductive method.

For a student devoted to physical science though, as the gifted mathematicians Poincaré and Maxwell have been anxiously insistent that he should be aware, there are lurking elements of danger in magnifying a bare logical skeleton as a goal; and in spending best effort upon articulating it. It is a misguidance apt to control into rigidity thought which can scarcely prove worthily fruitful unless it is maintained plastic. There is a plain sense in which dependence upon clarity of demonstration in terms of mathematical brevity and rigor may operate as a defect; and that severe pruning which suppresses all but definitive advance may mislead. There is a season for mitigating the austerity of algebra and daring to become discursive, for relaxing the ambition that is steadied to attain command of abstract principles on their highest level and for pausing in reflective examination of their genesis and their setting. Truly it would sterilize action to incline thus always; but never to turn aside from the more arduous pursuit tends to dissipate that atmosphere for dynamics which has given it life.

At the other extreme are found the practical temperaments, looking for tools with which to undertake their special tasks, and largely unmindful of the processes by which those have been shaped and of the far-reaching equipment in which their function is but one part, if only a particular routine can be adequately served or intelligently mastered. And this more empirical frame of mind that springs from absorption in monopolizing pursuits can be fostered and strengthened by the sheer difficulties in external form that are impressed upon abstract dynamics by the tendencies that have just been referred to, and by the air of remoteness from things material and mundane which that treatment, if uncorrected, confers. Unless it can be halted, therefore, a movement toward disintegration which must be coped with will confront the cultivators of dynamics that derives a backing also from other circumstances of the present situation.

The lifting of technical science to a better plane, where the habitual facing of new problems under the illumination of theoretical insight is coming to prevail, creates a demand in all the fundamental sciences that is a modern appeal. It has been incorporated into fixed plans of preparation for normal careers in active life to accomplish those things which were formerly undertaken with dominating inclination by minds self-selected through their special gifts. There must be, then, in the methods of presentation and in the execution of them, some recognition of a constituency that is at once larger, less homogeneous, and more in need of aid. In a restricted sense of the word, there is a summons to popularize the abstruser sciences, and among them dynamics, with a design to favor their assimilation by students at an earlier stage. This will make concessions in view of hindrances inherent in the subject-matter, and allowance for faculties of comparison and of analytic judgment not yet ripened into full command of all resources.

There is some element in the immediate need that is due to passing a transition and that will be lost in a newly adjusted order; for it has appeared from manifold experience what marvels can be wrought by tradition in giving easy currency to scientific doctrine. Moreover, the obstacles that loomed larger by mere novelty suffer genuine reduction by more lucid statement. An older generation arrived but gradually at an understanding of the principle in conservation of energy, and caught the advantage and power of absolute measurements first in glimpses. Yet they have lived to find those unfamiliar ideas adopted among the smoothly working formulas of unquestioned truth. So it will not pass the limits of a reasonable anticipation to forecast how the younger generation of today can move at ease in their maturity among bold concepts that were obscure when imperfectly grasped. Nevertheless, as the call now is, so must the answer be given.

Every aspect of the thoughts here put down is framed in a personal experience: the profit from quickening perception and appreciation for the nexus between sharply generalized ideas and their narrower origins; the benefit of laying stepping-stones gauged to a student's stride; the reward of implanting human interest within the routine of an industrial calling; also the moral gain through confirming intellectual honesty under a sustained demand for actual comprehension of what one is challenged to attack among the papers rated as classics, or in judging and sifting recent work. Aiding to scent difficulties first and then to overcome them fits the processes of the average mind, where the stronger talent can walk self-guided.

The present enterprise was born of the foregoing considerations in so far as they dictated its material and the ends for which that was offered in gradual accumulation during many years and under the influence of contact with students of varied purpose. It renounces from the outset all claim to be systematically conceived; it is content with a circling return from one point and another to a core of ideas that are worth reviewing in their various aspects because they are central. In their nature being a supplement to standard books that differ in type from each other, and offering themselves in flexible continuation of an elementary stage with unsettled achievement, these selected discussions cannot escape being judged fragmentary by some, redundant by others. But their spirit and their general aim are built upon ascertained failure to acquire elsewhere a just comprehension of several matters here made prominent and perhaps in some degree originally presented.

This kernel of intention in the subject-matter gathered for these chapters lends to them, it may be claimed legitimately, something of peculiar appropriateness for the circumstances of their publication. On the occasion to be celebrated it seems particularly pertinent that there should be recorded in some permanent form the working of those influences which our University has not withheld from her graduates, to nourish in them a living root of independent thinking and of unflinching thoroughness without which constructive scholarship cannot exist.

June, 1917.

CONTENTS

CHAPTER I

Introductory Summary

1. Only sciences that have attained a certain ripeness of strongly rooted development have been found capable of combining a vigorous and progressive activity at their working frontier for advance with reflective examination of their deeper foundations and their general method. The activity is aggressive in devising novel attack upon enlarging material, while reflection upon what has already become standard must go with recasting it to meet modified demands. This situation has been prominently realized in the case of dynamics, whose stirrings to self-criticism have been evermore spurred by the interactions with mathematics and astronomy, its closer neighbors, at the same time that its field was broadening to permeate and harmonize the greater part of physics. A large net gain of helpful stimulus from common aim must be allowed here, reënforcing the vigor from rapid growth, though there have been some dangers for dynamics to avoid, such as becoming infected with the more formal and abstract spirit of mathematics, or underrating its own basis in phenomena by acquiescing too generously in philosophy's rating for empirical science. It is a fitting preliminary to our immediate purpose to touch upon one or two such reactions between influences from without and from within; in part because the inquiries that were provoked, though prolonged through fifty years or more with acuteness and tenacity, have left practically unshaken the external forms of quantitative expression, at least. This is no sign, however, that dynamics is stationary and stereotyped; but only a reassuring fact to beget confidence in the fabric of the science. The subtle and less obtrusive changes

must not be forgotten, that have clarified the concepts and infused into them added significance by revised interpretation. Reading the prospects of the imminent future, too, rouses the expectation that what has been will continue to be, while dynamics is adapting itself to a wider scheme of connections and to a more accurate insight into its own doctrine or theory.

It is indeed an astonishing testimony to the happy strokes of genius in the founders of mechanics that force, impulse, work, momentum and kinetic energy still exhaust the primary needs, though the broader scope of dynamics now covers the chain of transformations in which mechanical energy is only one link. And it confirms our belief in the vital and definitive appropriateness of those quantities to find them retained essentially by those who are trying out another body of principles that might be substituted entirely or in part for the Newtonian mechanics. Meanwhile the equations of motion have not been superseded, yet they date from the seventeenth century; the notable advances due to d'Alembert, Euler and Lagrange in the eighteenth century, and to Hamilton in 1835, offer still the foundations upon which we build. But this introduction would outline a one-sided and misleading picture of mere static stability unless it used its warrant in bringing to supplementary notice three strands that have been woven into dynamics more recently, to alter in some degree its texture and to influence its emphasis. We shall next attempt to dispose of these in all proper brevity.

2. Under the first label *energetics* we are called upon to chronicle a strong movement that sought to enhance the prestige that energy in its various forms had already gained by the rapidly successful campaign about the middle of the nineteenth century.[1] This tendency was an almost inevitable accompaniment of that dominating relation to physical processes which conservation of energy as a conceded central principle had justified beyond cavil.

[1] See Note 1. Refer to collected notes following Chapter IV.

But the more pronounced utterances about energy overshot the mark in their zeal, and sought to exalt it in rank as the one dynamical quantity to which the rest should be held auxiliary, and upon which they should be based mathematically. Then the series—kinetic energy, momentum, force, mass—was to be unfolded out of its first term by divisions; and violent extremists were heard, even condemning force as a superfluous concept, refusing to associate it directly with our muscular sense, or to recognize it as an alternative point of departure yielding momentum and other quantities by multiplications. Of course deliberate minds looked askance at a professedly universal point of view that would exclude, save at the cost of an artificial device, such important elements as constraints that do no work. Common sense declined to cripple our assault upon problems for doctrinaire reasons that would bar and mark for disuse certain highways of approach, but it seized the chance instead to enrich and strengthen dynamics by wisely adopting the suggestion to exploit more completely the relations that energy specially furnishes, and to incorporate them among its resources and methods. After abating its flare of exuberance, the saner forces behind the reconstruction that was advocated have been harnessed and made contributory to a real advance that grafts new upon old, and embraces whatever proved advantage attaches to all reasonable points of view, with the object of reducing finally their oppositions and fitting them in place within a more comprehensive survey.

What is patent to read in the example of energetics should in prudence be made further to bear fruit; since judging historically, any new burst of reform spirit will be likely to repeat the main features of its lesson. An old and thoroughly tested science especially will less easily break the continuity of its course, though it is always responsively ready to swerve under every fresh impulse to amendment by discovery. So the matters offered recently under the caption *relativity* are surely giving to dynamics a wider sweep of horizon; but there too, when the permanent benefit accruing has been sifted out, the residue will probably prove more moderate than the tone of radical spokesmen has been implying while the sensation of novelty was strongest.[1]

3. It has been remarked often that Newton's three laws of motion taken by themselves give a bias toward concentrating attention upon momentum, and upon force exclusively as its time-derivative, with a comparative neglect of the counterpart in work and its relation to force. The restoration of balance began at once however, and soon the *principle of vis viva* was added and recognized as complementary on a level footing to Newton's second law. The equivalents of what are now known as the impulse equation and the work equation were established firmly and put to use. The readjustment thus begun was continued by steps as their desirableness was felt until with the ripeness of time it culminated, we may say, in the proposals that form the nucleus of what we call energetics. It will be profitable to expand that thought and mention some chief sources of the need to follow that line, or what gain has been found in doing so.

In rudimentary shape the idea of conservation of energy had emerged early; the histories are apt to date it from the method invented by Huyghens for the treatment of the pendulum. And so soon as the formal step had been taken in addition, that set apart under the heading *potential energy* the work of weight and of gravitation, because it can be anticipated by advance calculation exactly and with full security, the invariance of mechanical energy under the play of these forces when thus expressed, or its conservation within these narrower limits, became a demonstrable corollary of fundamental definitions.

[1] See Note 2.

The discovered inclusion of electric and magnetic attractions or repulsions under the same differentially applied *law of inverse square* that is characteristic of gravitation made natural the extension of potential energy as a statement of securely anticipated work to the field of those actions as well. And a large group of valuable mathematical consequences was accumulated which remain classic and which accompany the law of inverse square wherever it may lead, retaining their validity with only slight changes of detail.

These developments are controlled to a great extent by the idea of energy, and they must have built up a general perception of its power. The invariance of energy was fitted more completely for use as a principle, wherever its mechanical forms alone enter which we distinguish as kinetic and potential, when Gauss had evolved that plan of so-called *absolute measure* which has furnished us with the centimeter-gram-second system. He certainly consolidated into unity all sources of *ponderomotive force* in the several fields where a potential had been recognized. Of course we discriminate between this stage and the conservation of energy under all its transformations to which the period of Mayer, Joule and their coworkers attained. The earlier halting-place behind distinct limitations of scope left matters besides with a formal content only, in the sense that no questions were raised and squarely faced that looked toward localizing the latent energy and investigating the possible mechanism by which a medium might hold it in storage. This formal mathematics centered on the fact that the work done within a *conservative system* and between the same terminal configurations does not depend upon the particular paths connecting them. It is a strikingly significant exhibition of that quasi-neutrality that is now one salient and accepted feature in the procedure of energetics that so much of solid and permanent accomplishment was possible while certain vital issues were evaded, and without being compelled to register even a tentative decision upon them. That *non-committal* attitude towards much else as subsidiary, provided always that the gains and losses of energy for the system under consideration can be made to balance, has often been employed to turn the flank of obstacles and has been in that respect an element of strength. Or it leaves us in the lurch weakly, we might say about other occasions where we have stood in need of some crucial test between alternatives, and have found but a dumb oracle.

4. The next important advance was then timely and specially fruitful in giving life and deeper meaning to what had been in these directions more a superficial form; and at the same time in moving forward beyond the previous stopping-place to expand the range of dynamical ideas.[1] It is Maxwell who is credited with initiating these contributions by treating dynamically new aspects of electromagnetic phenomena. He took bold and novel ground by outlining his provisional basis for an electromagnetic theory of light that converted a colorless temporary vanishing of energy into a definite and plausible plan for its storage in a medium. In achieving this change of front he brought three lines of thought to a convergence-point; for besides the researches of Faraday and those that identified quantitatively the many transformations of energy, he utilized more fully than his predecessors had dared the possibilities that the earlier dynamics had done much toward making ready to his hand. It is this third element perhaps that marks most strongly for us the threshold of the new enterprise upon which dynamics will hereafter be engaged, in whose tasks we can find a union in just proportion of imaginative speculation with mastery of the mathematical instruments and with the candid policy of energetics to preserve an open mind and a suspended judgment in the face of undecided questions.

[1] See Note 3.

Maxwell was a pioneer in prolonging with new purpose the sequence upon which d'Alembert set out, and which Lagrange continued, beyond the point at which the latter paused after recording notable progress. What those earlier men had done with the discovery of *virtual work* as a basis for developing mechanics remained to be restated for dynamics, and adapted to a more inclusive command of energy transformations. Among other things this has given us an enlarged interpretation of older terms. We are ready to view a conservative system as one whose energy processes are reversible: that is, energy of any form being put in, it can be restored without loss, in the same form or in some other. We have learned to group fair analogues of kinetic and of potential energy for a system thus conservative according to one defensible test. Potential forms of energy will be found *resilient* as the original examples are; that is, they will exhaust themselves automatically, under the conditions of the particular combination, unless the corresponding transformation is prevented actively. But in order to be coördinated with kinetic energy on the other hand, the passive quality must be in evidence that requires some decisive intervention for the passage into other forms. This trend toward assigning wider meaning to dynamical concepts has given us further *generalized force* as a quotient of energy by a change in its correlated coördinate; the matching of force and coördinate as factors in the product that is energy being executed on due physical grounds. We have been led likewise to replace mass by a broader term *inertia*, where a quantity is detectable in the phenomena of more general energy-storage, that stands in essential parallelism with the relation of mass itself to force and kinetic energy. And the dynamical scheme has been rounded out by allowing to momentum those privileges of latency and of reappearance in the literal mechanical form, that were at the outset the monopoly of energy.

5. These comments have been attached to Lagrange's equations because Maxwell did in fact make them the vehicle of his thought; insisting upon sufficient detail to lift the reproach of indefiniteness, but also by a right inherent in the method passing over in silence the points where invention had thus far failed. But it was demonstrated long ago that d'Alembert and Lagrange and Hamilton have provided us with interconnected lines of approach to the same goal; except as the element of choice is directed by convenience Hamilton's principle lends equal favor and support with Lagrange's equations to the attempt to summarize a comprehensive statement in terms of energy. The former however elects to generalize for all analogous transformations upon a simple theorem: That potential energy will exhaust itself as rapidly as imposed constraints allow upon producing kinetic energy.

Beside the direct intention to indicate some reasons why dynamics leans increasingly upon energy relations, and borrows from energetics some modes of attack, these later remarks have a reverse implication as well. They intimate the belief that firm hold upon the elementary content of dynamical principles and intelligent full insight into them are not superseded, nor yet to be slighted. And the meaning here is not the mere commonplace truth that the more modest range satisfies many needs; or that historically it is the tap-root that has nourished and sustained the later growth. But recurring to what lies at the foundation is further the best preparation for the critical discrimination that must be exercised at the advancing frontier, because it holds the clews of conscious intention by which all effort there has been guided, and lends effective aid in steering an undeflected course among a medley of proposals to tolerate in concepts a figurative shading of their literal acceptation, or to condone acknowledged fictions on grounds of expediency.

6. The redistribution of emphasis upon which we have been dwelling has doubtless exercised the most penetrating influence to alter the complexion of mechanics as Newton left it, and therefore we have put it first. But there has been a second movement whose modifying effect as dynamics has grown must not be neglected, and which also like the leavening with energetics has been spread over a considerable period, though our report of its outcome can be compressed into a brief space.[1] This exhibited itself in a searching and protracted discussion on the *relativeness of velocity and acceleration* that did its part in contributing to clearness by removing ambiguity from a group of terms and carrying through a completer analysis of their bearings. The main concern here was not so much with the baldly kinematical side of the question; since it is plain that the final truth in that sense lies very near the surface. But the endeavor was quite specially shaped by the ambition to contrive at least soundly consistent expression for all dynamical processes that shall be recognized in physics; perhaps with some reach toward an ideal of universal and ultimate validity. The entire relativeness of those motions, which furnish leading factors of importance in decisions upon working values of dynamical quantities, is now a standard item in the opening chapters of dynamics as a corollary to choice of reference elements by agreement.

The acquirement of this point of view has therefore excluded all search for truly *absolute motion* and canceled the unqualified significance of the phrase which dates as far back as Newton. Since it seems flatly contradictory to unshackled relativeness, an impression may be created at first hearing that here for once the older thought has been overturned and radically revised. Yet the case is not so weak as it sounds, nor do we see, when we look below the surface, that any foundations have been affected vitally. We may be comforted to observe only another striking instance where a great mind did not everywhere and straightway hit upon most felicitous terms to describe how it

[1] See Note 4.

dealt with powerful nascent conceptions. Newton seems to call motions absolute if they dovetailed easily with the spacious frame of physical action that his discovery of gravitation was beginning to build; and himself engrossed in a swift reconnaissance through the new region, he left later invention to amend his notation. But it is chiefly the philosophical connotations of his word *absolute* and not its unfitness in physics that have made it the center of futile controversy. Thus the idea that the older writers really had in mind when they spoke of absolute motion was scarcely different from one that continues to hold its ground and compels us still to separate two lines of inquiry. Because beyond the settlement of kinematical equivalences that is direct and simple since it is unhampered by any physical considerations, the questions of real difficulty remain unsettled to confront us. They have had a certain elusive character by involving a complicated and tentative estimate that must balance on the largest scale and through the whole range of physics net gain against loss in simplicity. What common background, as it were, of reference-elements is decipherable upon which the interplay of forces and of energies shall stand in simplest and most consistently detailed relief?

In consequence it has not been displaced as a tenet of orthodox dynamical doctrine that standards by which to judge of the energy, momentum and force that ought to appear in its accounts will not stand on a par if adopted at random, however interchangeable they have proved in passing upon rest, velocity and acceleration by the mathematical criteria in the more indifferent domain of kinematics. Dynamics has never hesitated to stigmatize apparent forces, for example, as spurious or fictitious in relation to its general procedure, and to revise its lists of rejections on due grounds derived from advance in knowledge and in method. The definitive resolution of uncertainties that affect reasonable decision for the questions here implied is still awaited;

of necessity that objective is not attainable conclusively while the surveys in the several provinces of physics remain both fragmentary and disconnected. Though it has been claimed indeed that secure foothold was being gained through reliance upon a reference to stellar arrangement in removing excrescences that showed by the light of its corrective tests.

7. The growing practice to designate that reference as *ultimate,* however, has not excluded a proper admission that its lines of specification were to be improved by whatever greater precision new discovery and analysis of it reveal definitely to be progress. And it is fairly probable that majority opinion was looking entirely in that direction for fresh landmarks until other prospects were opened with vigor in recent years. These depend upon a certain increase in freedom to retain functional forms when the time-variable is added to the coördinates and included in the group of quantities that are involved in the readjustment when a change of base in the reference is undertaken. This far-reaching proposal derived its original suggestion from optical phenomena peculiar to electromagnetism and in one sense exceptional; yet since it is the crux of this situation that a decision of universal application is sought, any unreconciled indications of alternative must be reckoned with, whereby two plans for attaining the maximum simplicity that is desired become divergent. The competitive *schemes of ultimate reference* cannot be weighed decisively before the ramifications of both have been traced everywhere in that detail which can afford a satisfactory conclusion through their final comparison. And for that the time does not seem ripe; especially as each thus far falls short of established universal quality by seeming to leave some combinations unreduced, or abnormal to its plan. It is therefore reassuring to our logical sense to note how the practically available devices of proximate reference persist and are neutral, save in the formulation of the limits upon which their steps of increasing precision may be declared to converge. For that their own framework is by spontaneous intention approximate can be conceded without discussion.

8. The contrast upon which we have been remarking, between an indecision toward many-phased equivalences and the evolution of preference among them is then one characteristic of the transition from kinematics to dynamics; that is, from a range fixed by mathematical conformity to a selection narrowed by physical meanings. We can proclaim a forward step in that direction when the allowable mathematical range has been plausibly delimited, as with the transverse wave of optics from Fresnel's wave-surface in crystals to recent descriptive spectroscopy; but it is the crown of attainment to master insight into the causes of the effects observed, or into their sources, or into their explanations, in whatever chosen terms the phrase may stand. This persistent effort to identify physical sequences with a mechanism, to link a series of phenomena by means of a mechanical interpretation, has absorbed its full quota of sanguine activity since Newton scored his early partial success with the propagation of sound. The record shows in the main that the harvest of reward for these attempts has continued into this later era, slackening somewhat of course by exhaustion of the material. Yet there has been, too, a baffling of the imagination in its task of dissecting the complicated workings of energy in less traceable manifestations by traveling on parallels to direct sense-experience. And again optics illustrates; but now is shown a kind of failure, both with the abandoned types of its theory and in its electromagnetic alliance.

Every move in bestowing thus upon dynamics the control of a larger domain has been healthy growth, keeping pace with progress in other directions; and always sufficiently safeguarded against speculative vagueness by bonds with the method of its beginnings. Wherever mechanical energy in ponderable masses exhibits itself in the actual chain of transformations, it gives a touchstone through the measurable quantities, like ponderomotive force, by which to try the conceived series for its validity or consistency.

There are assumed successions, however, in which mechanical energy is not directly in evidence though equivalents of it appear in amounts known by using the change-ratios. Suppose we translate the given facts or quantities and introduce mechanical energy fictitiously. We have been prone to incline our judgment of the original case according to the analogies of its artificial substitute, and accordingly to accept the assumptions of the former or to speak skeptically of its paradoxes. But in the puzzling region that we have just mentioned there may be written a hidden caution about the cogency of such transferred conclusions. The absence of mechanical energy from the transformations that do occur, as we are ready to suppose for light during transmission, or for a free electron with inertia and without mass but traversing an electromagnetic field, may be a contributory circumstance in precluding a *mechanical model* and in leaving us thus far in the twilight of kinematics, wrecked on obstacles of seeming internal contradiction. And to the extent to which this indicated possibility is entertained, the leverage of these unreduced phenomena will be diminished, to guide or to modify dynamical thought that discusses ponderable materials.

9. The third gain that we must bring forward is the improved formulation of dynamics by replacing the cartesian expansions with vector analysis, whenever general discussions and theorems are taken in hand, or indeed everywhere unless we are barred by the needs of detailed calculation to which the vector notation is not so well adapted. The direct influence here is confined to external forms, it is true; yet indirectly an undeniable effect will always be exerted to favor continuity in the presentation of reasoning, and to preserve with fewer breaks an intelligent orientation during extended developments. These advantages are felt already, and they will accrue perpetually as a natural accompaniment of increased compactness in stating relations and of accentuating resultants first, only passing on to their partial aspects where necessary. We should all lend our aid to banish the obscurities and the disguises inseparable from the older system of equation-triplets. The subdivision of the newer analysis that is known distinctively as vector algebra is standardized fairly to the point of rendering great help in dynamics, and adjustments to this specific use are perfecting. As regards the vector operators like gradient, curl and divergence, they are as yet far from establishment in full effectiveness, by unforced extension of their original relation to field-actions and abatement of its comparative abstruseness.

10. This introduction will distort the truth of its own words and convey an unbalanced false impression, unless our reading of it can be depended upon to counterpoise the omissions that have trimmed it to these succinct proportions. So it is well to make room at this point for a few sentences that bear upon maintaining a real perspective against the tendency of extreme compression. And first it must be realized that the personal careers of a small group of geniuses do not constitute scientific history. To mention one great man and to picture him advancing with long sure strides implies with scarce an exception a whole accompanying period active with sporadic anticipations of some larger swing; an epoch of transition busy with foreshadowings of a new alignment. One's own thought should always supply this current of perhaps unrecorded preparation for an impetus that has given enduring reputation to its standard-bearer. The moulding of dynamics therefore is not the merit of its master-builders alone; we must not ignore those who had an inconspicuous share in establishing and in perpetuating its governing traditions.

Then secondly it may prove misleading to speak exclusively

of changes and innovations, though some temporary aim compels that. So we should return to the thesis of our opening paragraphs and allow them a corrective weight: That the large body of principles acquired early for dynamics and since unquestioned has steadied its course. It has been capable of assimilating the material that we have chosen to mention more explicitly without sacrifice of comparative power to treat for example the mechanics of solids and fluids. The considerations derived within that older territory must hold their place in what now follows.

11. It will be helpful in the direction of forestalling verbal quibbles and of clearing the ground otherwise if we enter next upon an explanation of the usage that we shall adopt for a few convenient terms; and also proceed to indicate the general attitude chosen in which to approach mathematical physics, of which dynamics forms one part. It may be well to premise once for all that no such personal choice covers a mistaken endeavor to close a question that is regarded reasonably as open, and to silence dissenting opinion. But there is often a practical necessity for taking a definite position, where adherence to one view colors exposition; and thus it should be candidly announced, although the occasion is not appropriate for extended argument.

In accordance with the unavoidable compulsion to take up piecemeal the phenomena and the processes given by observation and experiment in the physical world, any particular problem of dynamics is obliged to concern itself with a solution obtained under recognized limitations. These exhibit themselves on one side in setting a boundary to the region within which the course of events shall be investigated. If we distinguish within such a boundary a part enclosed that is ponderable and a part that is imponderable, we shall apply those terms on a plain etymological basis; so that the ponderable contents have weight as evidenced by the balance and are subject to gravitation, while the imponderable contents are not thus detectable. We shall speak of the former also as masses or as *bodies.* The latter if not alluded to as free space are called the ether, or the medium, meaning the medium for the transmission of light and other electromagnetic action. It is assumed that the ether-medium has not mass in the sense just specified; but this does not deny to it the more inclusive quality of inertia in certain connections.

A distinction need not be always upheld between mechanics and dynamics; but where this is done the second name has the broader scope, in that it may bring both masses and medium under consideration, which comprise then a dynamical system rather than a mechanical one. By contrast the older branch, mechanics, attempts only to deal with masses grouped into one body, or into a system of bodies. We shall conceive a body to fill its volume continuously and therefore to be adapted in so far to expressing by means of an integral its total, either of mass or of any quantity that is a function of the mass-distribution. The conception behind the phrase *system of bodies* is somewhat flexible; it may denote a discrete arrangement of bodies, whose mass and the like are then given as a sum of a finite number of terms, of which usage the astronomical view of our sun and its planets grouped as bodies in the solar system affords a typical instance; but it is applied also to a closely articulated assemblage of bodies like a machine, under suppositions that might or might not naturally justify integration. The opposition between body and system of bodies is retained and does some service though it is not tenable under stricter scrutiny, and cannot be radical so long as physical theory actually analyzes all accessible bodies into fine-grained systems for the purposes of molecular and atomic dynamics. On the other hand the contrast between systems of bodies and dynamical systems loses somewhat in significance where the interspaces are assumed to be void and the ether-medium is ignored; an abstraction common everywhere but in electro-magnetism; and the epithet, dynamical, then points only towards inclusion of all transformations of energy that remains associated with masses.

12. The tangled complexity in phenomena as they occur however compels our official accounts of them to be given piecemeal in other respects than by isolation of the region that lies within an assigned boundary. What is further to be done may be denominated variously; but it runs toward idealizing conditions, both by selecting certain elements as most important for study of their quantitative consequences and by a restatement of these that consciously relaxes somewhat precision of correspondence with the facts. It is evident how the two sources of distortion are likely to conspire in simplifying the mathematics; since neglecting weaker influences puts aside their smaller effects as mere modifying terms of a main result. To prune difficulties by this procedure as a preliminary to formulation and discussion is in some sort a contrivance of approximation, conceding the lack of desirable full power in our mathematical machinery. That several determining reasons blend in it can perhaps be recognized, though that is a subtle question upon which we shall not touch; but what has practical weight is to separate two uses of approximation, if such omission be accepted as one of them, at the same time granting that both are drawn upon partly because mathematics limps.[1]

To put the case briefly, sometimes we lay down a rule strictly but approximate to the results of it; which is a purely mathematical operation, utilizing for example a convergent series as we do when calculating the *correction for amplitude* in the period of a weight pendulum. Or again the assumed rule itself is known to be approximate, as is the fact when we call the pendulum rigid and the local weight-field uniform and constant. A further

[1] See Note 5.

distinction is that the first type relates to obstacles which may be overcome entirely by device, as in reducing finally some obstinate integral, but which lie off the track of advances in physics. In the instance just quoted the correction for amplitude will remain untouched, because an angle and its sine will never be equal. But with approximations of the second or physical type it is otherwise; we cannot make a body more accurately rigid by taking thought, nor can we bestow upon the field-vector (**g**) any quality of constancy that it lacks; so they progress by changing their rule. If provisional and marking imperfect knowledge while we await amendments of magnitude not yet ascertained, they move toward refinement of precision parallel to the advancing front of experimental research, as the law of Van der Waals about gases is seen to improve upon that of Boyle. Yet no supreme obligation is felt to make such changes everywhere; permanent and voluntary renunciations of achievable accuracy are frequent, too; we shall probably continue in many connections to discuss rigid solids and ideal fluids, notwithstanding the volume of fruitful investigations in elasticity, in viscosity and elsewhere, whose data are now at our disposal.

13. All these points are self-evident at first contact, and yet it is advisable to name them, in order to put aside what is incidental and focus attention upon the intrinsic structure of our equations, which leaves them inevitably approximate as an accepted limitation due to idealized or simplified statement. Clothing this thought in a figure, let us say that the principles of physics crystallize from the data of discovery into the concepts that have been shaped by invention to express them, but not without revealing traces of constraint and distortion that are not subdued and made quite to vanish under repeated attempts at adjustment. Historical inquiry has brought to light some remarkable interdependences here, and furnished a list of examples how discovery has stimulated the invention of concepts to match, and how on the other hand a stroke of inspiration in devising a well-adapted concept has smoothed the path to discovery of principle. Nevertheless the intimate psychology of such reciprocity is one of those deep secrets that have been securely guarded, and it need not concern us; we reach the kernel of the matter for the present connection when we insist upon the framework of dynamics as built of invented concepts and add one or two corollaries of that central idea.

In the first place, in order to proceed by mathematical reasoning from specified assumptions, the margin of ambiguity in the terms that are used must be cut down as much as is feasible. A controversy about Newton's third law; whether or not it applies to a source of light, could be settled easily under our agreement that the ether-medium is not a body (*corpus*). And the emancipation from corroborative tests in the free realm of concepts is some compensation for the trouble of defining. It has been laborious to disentangle the mean solar second as a uniform standard of time; but the fluxion-time (t) of Newton in its quality of independent variable must be equicrescent. So in the concept of unaccelerated translation there is no place for differences of velocity anywhere or at any time; and values specified to be simultaneous cannot be affected by uncertain deviations from that assumption; and for the conceptual isotropic solid under Hooke's Law the stress-strain relation is rigorously linear. Likewise, if according to the tenets of relativity the light-speed in free space and relative to the source is always the same, we go on unflinching to work out the consequences; and any such assumption with its demonstrable deductions will be entertained with candor, so long as its contacts with observed facts given by correct mathematics do not fail either as plausible physics. However, from the side of these perpetual tests there is sleepless critical judgment upon all our mental devices, to continue, to revise or to reject them. In other respects the schemes may be plastic to shift the point in precision at which they halt, and we are reasonably tolerant also of conventional fictions.

This brings to a close the short preface of such verbal comment as may provide a setting in which to frame the equations that follow, and at the same time assist in some respects to receive more appreciatively their meaning by bringing to view what underlies them.

CHAPTER II

THE FUNDAMENTAL EQUATIONS

14. Any standard exposition of dynamics, though it may not attempt a comprehensive and most general treatment of the methods and principles, will introduce into its resources for carrying on the discussion the six quantities: Force, Momentum, Kinetic Energy, Power, Force-moment and Moment of Momentum. The terms in detail that are required for the specification of these, and a certain group of propositions into which they enter, are so fundamental that they become practically indispensable in establishing the necessary developments. The units that their function as measured quantities demands are supplied according to the centimeter-gram-second system with so nearly universal adoption that we can regard it as having displaced all competitors, everywhere except in some technical applications where special needs prevail; so that we shall consider no alternative plan of measurement.

Since the six quantities named are not independent of each other, but are connected by a number of cross-relations that we can assume to be familiar in their elementary announcement, it is clear that the way lies open to select for a starting-point a certain set as primary, the others then falling into their own place as derived or even auxiliary quantities. It is also plain, as a mere matter of logical arrangement, that any particular selection of a primary set will not be unique, with a monopoly of that title to be put first; and this leaves the exercise of preference to be governed ultimately by reasons drawn from the subject-matter. Not only is it possible to make beginnings from more points than one in presenting the six quantities on a definite basis, and in exhibiting the links among them, but it is the truth that beginnings have been made differently and defended vigorously. We have already alluded to one such period of polemic through which dynamics has passed. It is a necessity however to choose a procedure by some one line of advance; but let it be understood that we do this with no excessive claim for its preponderant advantage or convenience, and explicitly without prejudice to the validity of some other sequence that may be preferred.

15. In the light of this last remark we shall make our start by picking out for first mention a group of three quantities: Momentum, Kinetic Energy and Moment of Momentum. Without anticipating a more specific analysis of them, it is evident on the surface that they all apply in designating an instantaneous state depending on velocities, and that momentum is the core of the three; entering as free vector, as localized vector, and as factor in a scalar product. And further it can be noted at once, without presuming more than a first acquaintance with mechanics, that the remaining three quantities constituting a second group can be described in symmetrical relation to the first three as their time-rates. Then force is made central; and it in turn appears as a free vector, as a localized vector, and as a factor in a scalar product. We take the first step accordingly by laying down for application to any body or to any system of bodies the three defining equations:

$$\text{Total momentum} \equiv \Sigma \int_{\mathrm{m}} \mathbf{v}\,\mathrm{dm} \equiv \mathbf{Q}; \tag{I}$$

$$\text{Total kinetic energy} \equiv \Sigma \int_{\mathrm{m}} \tfrac{1}{2}(\mathbf{v}\cdot\mathbf{v}\,\mathrm{dm}) \equiv \mathrm{E}; \tag{II}$$

$$\text{Total moment of momentum} \equiv \Sigma \int_{\mathrm{m}} (\mathbf{r}\times\mathbf{v}\,\mathrm{dm}) \equiv \mathbf{H}. \tag{III}$$

These indicate in each case, with notation that is so nearly standard as to carry its own explanation, the result of a mass-summation extended to contributions from all the mass included in the system at the epoch, under the terms of some agreement covering the particular matter in hand, and isolating in thought temporarily, for purposes of study and discussion, the phenomena in a limited region. In conformity with a previous explanation in section 11 any assumed continuous distributions of mass are included under the integrals, whose further summation indicated by (Σ) may be necessary when a system of bodies, discrete or contiguous, is to be considered. It deserves to be emphasized perhaps that these are defining equalities merely; so that ($\mathbf{Q}$) and ($\mathbf{H}$) and (E) only denote aggregate values associated with the system at the epoch, and so to speak observable in it; neither side of the equalities conveys any implication about *external sources*, or causes by whose action these aggregates may have originated, or which may be operative at that epoch to bring about changes affecting them.

16. Because the variables ($\mathbf{r}$) and ($\mathbf{v}$) occur in the quantities with which we are now dealing, if for no deeper reason, it is implied that a definite system of reference has been fixed upon as an essential preliminary to actual attachment of values to momentum, kinetic energy and moment of momentum. For the ordinary routine which is likely to involve recasting vector statements into semi-cartesian equivalents, or the inverse operation of arriving at the former by means of the latter, the requisite elements for the reference are obtained by selecting an origin from which to measure distances and axes for orienting directions. Unless special exception be explicitly noted we shall follow the prevalent usage of taking axes of reference that are orthogonal and in the cycle of the right-handed screw; and shall for convenience conduct the main discussion on this permanent background, reserving any substitution of equivalents for occasions where that has some peculiar fitness. The reference-frame that has been agreed upon, it must not be forgotten, is in the essence of it *conceptually fixed* while the agreement to use it continues in force, because it has been singled out as the unique standard in relation to which we specify or trace what can be called the configurations (**r**) and the motions (**v**).

As an antecedent condition of algebraic evaluation for our three fundamental quantities in a given system at any epoch, the choice of some reference-frame then is necessary; but it is likewise evident that any one choice that may be made is equally sufficient in respect to removing mathematical indeterminateness. And consequently it will be found true that much can be done in advancing a satisfactory exposition of dynamical principles to the point where we stand at the threshold of calculations that rest on a basis of observed phenomena, without going beyond the potential assumption of that reference-frame that must be faced finally, in order to complete the necessary and sufficient condition for the definiteness of the physical specifications. In other words, a considerable proportion of the usual developments in dynamics can be provided ready-made to this extent, and yet fitting the measure of any reference-frame that is particularly indicated as appropriate by a physical combination or by a line of argument.[1]

These considerations are adapted to bring to the front also the idea that quantities like the three with which we are concerned at this moment can be evaluated for two or more different reference-frames, perhaps with the object of reviewing their comparative merit, especially in being adjusted to the preferences of consistent physical views (see section 7). It follows naturally therefore that provision must be made quantitatively for transfers of base from one reference-frame to another, either in progress toward ultimate reference, as in abandoning a frame fixed relatively to the rotating earth, or as a device of ingenuity in order to reach certain ends simply. The material of Chapter III in large part bears upon questions of that nature.

17. The range of the mass-summations that are stipulated in

[1] See Note 6.

the expressions with which we are dealing can vary with time for several reasons that can be operative separately or concurrently. It is compatible with many conditions about boundary-surface that material may be added or lost, as is the case when gas is pumped into a tank or out of it, or when unit volume of an elastic solid gains or loses by compression or extension. Or it may fit the circumstances best to mark off a boundary that changes with time, as when we take up mechanical problems like those of a growing raindrop or a falling avalanche. The values of (**Q**, E, **H**) are accommodated to any complication of such conditions, with the single caution that the total mass shall then be delimited as an instantaneous state at the epoch.

We go on to assume, however, in connection with any transfers of reference that we are called upon to execute, that mass remains unaffected thereby in its differential elements and in its total, being guided by the absence of experimental evidence that mass, in our adopted use of the word, needs to be made dependent upon position or velocity. Assembling these suppositions, we see that mass will play its part in the equations as a pure scalar and positive constant, except as accretions or losses of recognizable portions may be a feature of the treatment. And consequently equation (II) can be made algebraic at once, since the vector factors are codirectional, and be given the form

$$\mathrm{E} = \Sigma \int_{\mathrm{m}}(\tfrac{1}{2}\mathrm{v}^2\mathrm{dm}), \qquad (1)$$

although the original model should be preserved besides, as a point of departure for parallelisms that will show themselves later.

18. Return now to examine the two remaining equations, in order to extract some additional particulars of their meaning. In the first the total momentum appears as a vector sum, so built up that its constituents are usually described as free vectors. This term is seen to justify combining the dispersed elements to one resultant, on reflecting that the predicated freedom of such vectors lies wholly in the non-effect of mere shift to another base-point; and that this renders legitimate the indefinite repetition of the parallelogram construction for intersecting vectors until all the differential elements have been absorbed into the total aggregate. But this incidental and as it were graphical convenience must not lead us to neglect the fact that we are nevertheless retaining the idea of momentum as a *distributed vector*, and continuing to associate each element of it locally with some element of mass. However formed its total belongs to the system as a whole; and it can be localized, as it sometimes is at the center of mass of the system, only by virtue of a convention or an equivalence.[1]

We can call the total momentum a free vector, of course; but its freedom does not quite consist in an indifference about its base-point; more nearly it expresses the inherent contradiction there would be in localizing anywhere what in fact is still conceived to pervade the mass of the system. At several points we shall discover how the service of vectors in physics makes desirable some addition to the formal mathematical handling of them. It will not be overlooked, finally, how the above analysis of composition enlarges upon the addition of parallel forces to constitute a total, through the similar properties of an algebraic and a geometric sum; the latter reduces to its resultant by complete cancellation in a plane perpendicular to the resultant.

19. In the third equation each local element of momentum has the attribute of a localized vector through definite assignment to the extremity of its radius-vector. It is not apparent that the vector product in which it is a factor is thereby determined to be unequivocally localized; but here again physical considerations enter that are extraneous to the mathematics; the practice tacitly followed localizes the several elements of moment of momentum, not at the differential masses to which they in

[1] See Note 7.

one sense belong, but at the origin in acknowledgment of their intimate connection with rotations about axes there, and of the origin's importance in determining the lever-arms when the mass-arrangement is given. Each differential moment of momentum thus located being perpendicular to the plane of its (**r**) and (**v**) of that epoch, is evidently also normal to the plane containing *consecutive positions* of the radius-vector; that is, (d**H**) is colinear with ($d\boldsymbol{\gamma}$), if the latter denotes the resultant element of angle-vector that (**r**) is then describing; and on this we can found a transformation that is worth noting. If (d**s**) is the element of path for (dm),

$$d\boldsymbol{\gamma} = \frac{1}{r^2}(\mathbf{r} \times d\mathbf{s}); \qquad \dot{\boldsymbol{\gamma}} = \frac{1}{r^2}(\mathbf{r} \times \mathbf{v}); \qquad d\mathbf{H} = \dot{\boldsymbol{\gamma}}(r^2 dm); \quad (2)$$

and the last equation reproduces differentially the type of an elementary and partial relation among moment of momentum, moment of inertia and angular velocity for a rigid solid. Only ($\dot{\boldsymbol{\gamma}}$) is here individually determined in magnitude and in direction for each (dm); no common angular velocity and collective moment of inertia are assigned, as they are in the case of a rigid solid, but with disturbance in general of the colinearity shown by (d**H**) and ($\dot{\boldsymbol{\gamma}}$) into a divergence of the resultant vectors for angular velocity and moment of momentum.

20. The three equations of section 15 are simplified remarkably whenever the condition prevails that the velocity (**v**) has a common value throughout the system that is in question. This state of affairs is designated as *translation of the system;* it may persist during a finite interval of time, or it may appear only instantaneously, and in either case naturally it entails a corresponding quality in the simplifications. When the condition of translation persists the common velocity (**v**) need not be constant; but the simultaneous velocities everywhere must be equal. The resulting forms applying to translation are then seen to be for a total mass (m),

$$\mathbf{Q} = \mathbf{v}\Sigma \int_m dm = m\mathbf{v}; \tag{3}$$

$$E = \tfrac{1}{2}(\mathbf{v}\cdot\mathbf{v})\Sigma \int_m dm = \tfrac{1}{2}mv^2; \tag{4}$$

$$\mathbf{H} = (\Sigma \int_m \mathbf{r}dm) \times \mathbf{v} = \bar{\mathbf{r}} \times m\mathbf{v}. \tag{5}$$

The last equation introduces the familiar mean vector ($\bar{\mathbf{r}}$) which locates the center of mass of the system through the mass-average of the individual radius-vectors ($\mathbf{r}$) according to the defining equation

$$m\bar{\mathbf{r}} \equiv \Sigma \int_m \mathbf{r}dm. \tag{6}$$

The last group of equations contains the suggestion from which has been worked out a notion that has had some vogue and convenience in dynamics: that of an *equivalent or representative particle* to which are attributed negligible dimensions but also the total mass, momentum and kinetic energy of the system. Equations (3, 4, 5) show that such a fictitious particle at the position of the center of mass of the system would replace the latter in respect of ($\mathbf{Q}$, E, $\mathbf{H}$) while translation continues. And since it is their ratio to other lengths that settles whether dimensions are physically negligible, the absurdity that there would be in concentrating momentum and energy into a mathematical point is sensibly mitigated.

21. Even when the condition is not met that simultaneous velocities shall be equal everywhere, a constituent translation can be carved out artificially from the actual totals ($\mathbf{Q}$, E, $\mathbf{H}$) at the epoch and for the system. Let every local velocity ($\mathbf{v}$) be split into two components in conformity with the relation

$$\mathbf{v} = \mathbf{c} + \mathbf{u}, \tag{7}$$

in which ($\mathbf{c}$) is assigned at will, but taken everywhere equal, and ($\mathbf{u}$) denotes whatever remains of ($\mathbf{v}$). Then substitution in the fundamental equations of section 15 will segregate the totals into a part that corresponds to translation and a supplement. Among the indefinite number of possibilities, we select one particularly fruitful plan for illustration. Let $(\bar{\mathbf{v}})$ be the mass-average of velocities determined by the condition

$$m\bar{\mathbf{v}} = \Sigma \int_m \mathbf{v}dm. \tag{8}$$

Then if

$$\mathbf{v} = \bar{\mathbf{v}} + \mathbf{u}, \qquad \Sigma \int_m \mathbf{u}dm = 0 \text{ necessarily}. \tag{9}$$

But we have also, in consequence of equation (9),

$$E = \tfrac{1}{2}\Sigma \int_m(\bar{\mathbf{v}} + \mathbf{u})\cdot(\bar{\mathbf{v}} + \mathbf{u})dm = \tfrac{1}{2}m\bar{v}^2 + \tfrac{1}{2}\Sigma \int_m u^2dm. \tag{10}$$

And further,

$$\mathbf{H} = \Sigma \int_m[\mathbf{r} \times (\bar{\mathbf{v}} + \mathbf{u})dm] = (\Sigma \int_m \mathbf{r}dm) \times \bar{\mathbf{v}} + \Sigma \int_m(\mathbf{r} \times \mathbf{u}dm). \tag{11}$$

In order to reduce the last term place $\mathbf{r} = \bar{\mathbf{r}} + \mathbf{r}'$, so that $(\mathbf{r}')$ like $(\mathbf{u})$ is departure of the local value from the mean. Then finally

$$\mathbf{H} = (\bar{\mathbf{r}} \times \bar{\mathbf{v}}m) + \Sigma \int_m(\mathbf{r}' \times \mathbf{u}dm), \tag{12}$$

in which the segregation according to mean values and departures from them is complete.

Taking equation (8) in conjunction with the first terms on the right-hand of equations (10) and (12), the idea of a particle at the position of the center of mass reappears, having the total mass (m), the total momentum $(m\bar{\mathbf{v}} = \mathbf{Q})$, and the kinetic energy $(\tfrac{1}{2}m\bar{v}^2)$. But whereas equations (3, 4, 5) covered the data completely, this contrived and partial translation with the mean velocity $(\bar{\mathbf{v}})$ leaves residual amounts of kinetic energy and moment of momentum; and these are due to departures from the mean values of $(\mathbf{r})$ and $(\mathbf{v})$, as the last terms in equations (10) and (12) indicate plainly, which items, as is also evident, have no resultant influence on the momentum. It is clear that this plan of partition is adapted to accurate use; but it proves to have some advantages too as the basis of an approximation, where the residual terms are in small ratio to the translation-quantities and can be neglected in comparison with the latter. The so-called *simple pendulum* affords one instance.

22. The recognition of elements of momentum as localized vectors brings in an additional detail of their physical specification; so this alone could be alleged as one valid reason for conceding to moment of momentum its place in the general foundation of dynamics. But we are now in a position to realize another advantage of which that third equation gives us control. Mean values have admitted elements of strength in smoothing out accidental or systematic differences in a series of data, and in enabling us to convert an integral into a product of finite factors. Yet this acceptable aid may be offset in part by such elimination on the whole of departures from the mean as is shown in

$$\Sigma \int_m \mathbf{r}' dm = 0; \qquad \Sigma \int_m \mathbf{u} dm = 0.$$

Now first inspection of equation (12) shows how it serves to retrieve by means of the vector products the divergencies that would be lost from sight in the mean values, and thus to piece out the support in that direction which equation (10) accomplishes through its scalar products, wherever we have an interest to gauge effects of divergence that are cumulative and not self-cancelling.

23. Before passing on to another topic it is worth taking occasion to remark that the values for the totals of momentum, kinetic energy and moment of momentum can be adjusted without difficulty to expression as summations extended over a volume; for in terms of the local density (δ) and element of volume (dV) the mass-element there is expressible by

$$dm = \delta dV.$$

This density will be rated always a pure scalar on account of its correlation with mass, and both density and volume are best standardized in dynamics as positive factors in the positive product that is mass, though it is not advisable to brush aside too lightly the combinations that the character of volume as a pseudo-scalar permits. Since the value of the density is zero throughout the volume that is left unoccupied by the supposed distribution of mass, the inclusion of such portions into a summation throughout the whole region within the assumed boundary is without influence upon the result and can be indicated formally without error. To declare a density zero is the equivalent of excluding a volume from a mass-summation.

Hence the need of a double notation (Σ) and ($\int_m$) will disappear, if the continuous volume can be paired with a density also effectively continuous, by any of the plausible devices that evade abrupt changes at passage from volume with which mass is associated to volume from which mass is said to be absent.[1] With these words of explanation the alternative forms that follow are interpretable at once:

$$\mathbf{Q} = \int_{\text{vol.}} \mathbf{v}(\delta dV); \tag{13}$$

$$E = \tfrac{1}{2} \int_{\text{vol.}} \mathbf{v} \cdot \mathbf{v}(\delta dV); \tag{14}$$

$$\mathbf{H} = \int_{\text{vol.}} (\mathbf{r} \times \mathbf{v}(\delta dV)). \tag{15}$$

Let us add for its bearing upon the lines of treatment when mass is variable that then both (δ) and (dV) are susceptible of change. And also recall how there will always be that outstanding question about mean values in comparison with divergence from them, of which we spoke above, whenever we face the contradictory demands of mathematical continuity and of open molecular structure, in order to reconcile them adequately—for instance, in the concept of a homogeneous body with a value of density that is common to all its parts.

24. We shall next approach the remaining group of fundamental quantities that we have enumerated already as three: Force, Power, and Force-moment. The first object must be to set forth in satisfying clearness and completeness their relations

[1] See Note 8.

to the previous group of three, in order to proceed then securely with reading the lesson how the interlinked and consolidated set of six quantities provides all requisite solid and efficacious support, both for the current general reasoning of dynamics and for its specialized lines of employment.

We began to follow the track over ground that has become well-trodden since Newton's day, when we laid down a meaning for the phrase *total momentum of a system of bodies* and the symbol (**Q**) representing it which in effect only renames the intention of the historic words "Quantity of motion." We also continue the tradition that has been perpetuated ever since Newton's second law launched its beginnings for approval, by fixing attention in its turn upon the rate of change in the momentum, in its differential elements and in its total, and regard that as delivering to us the clews, that we shall later follow up, to the forces brought to bear upon the system of bodies that is under investigation, with which forces we must undertake to reckon. The gist of that law has not yielded perceptibly under all the proposals to improve upon it, though we may be rewording it more flexibly under widening appeal to experience. Its drift makes the claim that changes in (**Q**) are not *spontaneous;* that when they are identified to occur there is reason to be alert and detect why, with gain for physical science in prospect by success.

25. The first move toward formulation could scarcely be simpler; it is to indicate the time-derivative of equation (I) and write

$$\dot{\mathbf{Q}} \equiv \frac{d}{dt}[\Sigma \textstyle\int_m \mathbf{v}\,dm]. \qquad \text{(IV)}$$

Yet the mere mathematics of execution blocks the way with distinctions to be made, unless we are resolved to carry an overweight of hampering generality. For it is common knowledge that the masses are often comprised in such a summation on a justified footing that they are in every respect independent of time; and consequently it is then legitimate to differentiate behind the signs of summation in equation (IV). But forms alter as the mass included is in any way a function of time; they will differ besides if only the total mass changes by loss or gain of elements, or only the elements change leaving the total constant, or if both sorts of dependence upon time are permissible under the scheme of treatment. The first supposition of complete mass-constancy underlies the dynamics of rigid solids and is a stock condition in much dynamics of fluids as well. And because it prevails most naturally to that extent, it is perhaps fair to select this mass-constancy as standard; especially when departures from it are likely soon to be cut off from the stream of systematic development by running into specializing restrictions and a narrowly applicable result.[1]

However opinion may stand on that matter, it seems certain that no aspect should be allowed to escape us finally that belongs to the full scope of mathematical possibility attaching to the indicated time-derivative of (**Q**). Any contribution to the changes in momentum may mature a suggestion about force-action and gain physical meaning. Therefore the tendency seems unfortunate to borrow the terms of Newton's second law, for its professedly general statement, from the special though widely prevalent case which throws all the change in momentum upon the velocity-factor. To speak of force as universally measured by the product of mass and acceleration is misleading if the habit blinds us to the fuller scope of the second law, and atrophies at all our capacity to use it.

26. In order that the derivative of an expression may be formed for use, certain conditions of continuity must not be violated, as we know; but when a derivative is to be made representative of a sequence of states, mathematical physics has available a repertory of resources in constructing this requisite

[1] See Note 9.

continuity of duration and distribution. Examples are plentiful among the classic methods of attack, how variously the proper degree of identifiable quality is assigned to a succession of states, that links the individual terms into a continuous series. Rankine's device for studying a sound-wave in air is a travelling dynamics that keeps abreast of the propagation; Euler's hydrodynamical equations stand permanently at the same element of volume, and record for successive portions of liquid that stream by; and many processes where material passes steadily through a machine are most tractable in similar fashion. We shall not insist further then upon this point, except to say that advanced stages of the subject are less apt to rely upon straightforward sameness and constancy in the masses specified for summation under the term *body or system of bodies.* With the reserves of that cautious preamble, we can afford to qualify the case of mass-constancy and literal sameness as standard in a limited sense, and exploit some of the consequences flowing from that assumption.

27. On the grounds now announced explicitly the indicated operation of equation (IV) yields

$$\dot{\mathbf{Q}} = \Sigma \int_m \dot{\mathbf{v}} dm \equiv \Sigma \int_m d\mathbf{R} = \mathbf{R}. \tag{16}$$

As a symbol, therefore, (d**R**) is defined to mean the local resultant force at each differential mass for which there is evidence through the local acceleration; and accordingly (**R**) denotes the vector sum of such elements of force when the whole system of bodies is included. This *total force* is in nature a dispersed aggregate like the total momentum, and the line of comment under that heading applies here with a few changes, which however are obvious enough to absolve us from repeating it.

28. Before we carry the discussion into further detail it seems best to bring equations (II) and (III) to this same level by putting down their time-derivatives, observing consistently there also the imposed limitation to complete mass-constancy, but remembering always that we halted exactly on that line and postponed until due notice shall be given the further step in restriction that will introduce a rigidly unchanging arrangement or configuration of all the mass-elements. Writing first the general defining equation as preface,

$$P \equiv \frac{dE}{dt}, \tag{V}$$

we can then make the application to the specialized conditions that gives

$$\frac{dE}{dt} = \tfrac{1}{2}\Sigma \int_m \frac{d}{dt}(\mathbf{v}\cdot\mathbf{v}dm) = \Sigma \int_m(\mathbf{v}\cdot d\mathbf{R}). \tag{17}$$

This indicates at each element of mass a local manifestation of *power* that is measured by the scalar product of the force-element and velocity—this scalar product being of course not merely formal, since ($\mathbf{v}$) and ($d\mathbf{R}$) are not in general colinear. It has been called also, perhaps with equal appropriateness, the *activity of the force.*

29. In this preliminary consideration there remains only the time-derivative of equation (III). And we shall preserve a helpful symmetry of statement by giving its place here also to the general defining equation, and following it as we have done previously with its present special value. Then

$$\mathbf{M} \equiv \dot{\mathbf{H}}; \tag{VI}$$

and

$$\mathbf{M} = \Sigma \int_m \frac{d}{dt}(\mathbf{r} \times \mathbf{v}dm) = \Sigma \int_m(\mathbf{r} \times d\mathbf{R}); \tag{18}$$

the reduction of the expansion to one of its two terms being the evident consequence of the identity of ($\dot{\mathbf{r}}$) and ($\mathbf{v}$). The last equation demonstrates within the limits set for it that the time-derivative of the total moment of momentum measures the total force-moment of the local elements of force that are calculated according to equation (16). As a postscript to equation (18) repeat with the necessary modifications what was inserted in section 19, about equation (III), and at the end of section 22. The example of a *force-couple* will come to mind at once, where the pair of its forces is self-cancelling from the free-vector aggregate of force, and it devolves upon the localized force-vector of a moment to restore for consideration the important effects of couples.

Observe also the peculiar prominence of the radius-vector in vector algebra. Where the cartesian habit is to bring both moment of momentum and force-moment into direct and exclusive relation to a line or axis, vector methods substitute relation to the origin, which is a point. Upon examination, however, the difference partly vanishes, because the vector reference to a point is only a superficial feature. We have explained in connection with equation (2), how a resultant axis is tacitly added. The element ($d\mathbf{M}$) is similarly a maximum or resultant, the factors in ($\mathbf{r} \times d\mathbf{R}$) being given, the effective fraction of the moment for other axes through the origin being obtainable by projecting ($d\mathbf{M}$).[1]

30. Equations (16, 17, 18) bear on their face and for their particular setting sufficient reasons for interpreting ($\dot{\mathbf{Q}}$) in terms of those forces ($d\mathbf{R}$); (P) or (dE/dt) in terms of the activity of those same forces ($\mathbf{v} \cdot d\mathbf{R}$); and ($\mathbf{M}$) or ($\dot{\mathbf{H}}$) in terms of their force-moments ($\mathbf{r} \times d\mathbf{R}$). There seems to be neither confusion nor danger imminent if we extend the names thus rooted in commonplace experience to the (at least mathematically) more complicated possibilities of equations (IV), (V), (VI). We can be bold to identify ($\dot{\mathbf{Q}}$) always as some force ($\mathbf{R}$); (dE/dt) as a power (P); and ($\dot{\mathbf{H}}$) always as a force-moment ($\mathbf{M}$); if we have made ourselves safely aware how terms in any completed mathematical expansion may remain non-significant physically until

[1] See Note 10.

discovery confirms them. We have alluded before to the fact that dynamics does not altogether shrink from a figurative tinge in extensions of terms first assigned literally, if essentials of correspondence are adequately preserved. But notice particularly that the verbal adoptions proposed above cannot of themselves assure the occurrence of the duplicate adjustments among equations (16), (17), and (18). To forces whose sum is ($\dot{\mathbf{Q}}$) will correspond activities that we may denote by ($\mathbf{v}\cdot d\dot{\mathbf{Q}}$), and moments of type ($\mathbf{r} \times d\dot{\mathbf{Q}}$). But we must not conclude in advance that the former group will in their sum match (dE/dt); nor that the latter group will match exactly ($\dot{\mathbf{H}}$); though both equivalencies hold under the condition of mass-constancy. And for discrepancies there will be no general corrective formula; they must be newly weighed wherever they may appear.

31. Let us next turn back to the ideas of translation and equivalent particle of which we spoke in sections 20 and 21, and continue them in the light of equations (16, 17, 18). In the first place note that the mean velocity ($\bar{\mathbf{v}}$) as previously specified by equation (8) becomes now identical with the velocity of the center of mass, because the time-derivative of equation (6) takes the form

$$m\dot{\bar{\mathbf{r}}} = \Sigma \int_m \dot{\mathbf{r}} dm = m\bar{\mathbf{v}}. \tag{19}$$

Secondly the conditions justify for the next time-derivative,

$$m\dot{\bar{\mathbf{v}}} = \Sigma \int_m \dot{\mathbf{v}} dm, \tag{20}$$

showing that the center of mass has the mass-average of accelerations. Hence a particle having the total mass (m) of the system and retaining always its position ($\bar{\mathbf{r}}$) at the center of mass would show at every epoch the total momentum (**Q**); and its acceleration would determine the value (**R**) of the total force in equation (16) through the product ($m\dot{\bar{\mathbf{v}}}$).

But if the first terms in the second members of equations (10) and (12) and the derivatives of those terms with regard to time be now considered, with the new meaning for $(\bar{\mathbf{v}})$, it is seen that the specified particle at the center of mass duly represents all the dynamical quantities for the system, except those parts which depend upon departures $(\mathbf{r}')$ from the mean vector $(\bar{\mathbf{r}})$ and upon departures $(\mathbf{u})$ from the mean velocity $(\bar{\mathbf{v}})$. Hence such an artificial or fictitious *translation with the center of mass* runs like a plain thread through all the equations for the actual system, and reproduces accurately their six dynamical quantities when we simply superpose upon it the additional kinetic energy, moment of momentum, power and force-moment whose source is in the deviations from mean values. It is a self-evident corollary that in a real or *pure translation* the particle at the center of mass represents the system without corrections, since the local accelerations must be of common value while translation continues, in order that simultaneous velocities may remain equal. This keeps each velocity $(\mathbf{u})$ permanently at zero.

32. It will be instructive to enforce without delay the differences from parallelism with the preceding details that appear at several points, in the simplest combinations where it becomes natural to regard the total mass as variable with time. Let us then take up for consideration a body in translation, or equivalently a representative particle, denoting by (m) and $(\mathbf{v})$ the instantaneous values of its mass and velocity. For the momentum and the kinetic energy at the epoch we still have

$$\mathbf{Q} = \mathrm{m}\mathbf{v}; \qquad \mathrm{E} = \tfrac{1}{2}(\mathrm{m}\mathbf{v}\cdot\mathbf{v}). \tag{21}$$

If we stand by the agreement that $(\dot{\mathbf{Q}})$ shall be force and embody it in the time-derivative of the first equation, we shall write

$$\dot{\mathbf{Q}} = \mathbf{R} = \mathrm{m}\dot{\mathbf{v}} + \frac{\mathrm{dm}}{\mathrm{dt}}\mathbf{v}. \tag{22}$$

When mass is constant, resultant force and resultant acceleration have the same direction, as we can read in equation (16). But in striking contrast with that consequence, equation (22) shows that its (**R**) does not in general coincide with either velocity or acceleration.

Proceeding next to examine the power, and continuing to specify it as the derivative of (E) we find

$$\frac{dE}{dt} \equiv P = \tfrac{1}{2}(m\dot{\mathbf{v}}\cdot\mathbf{v} + m\mathbf{v}\cdot\dot{\mathbf{v}}) + \frac{1}{2}\frac{dm}{dt}(\mathbf{v}\cdot\mathbf{v})$$

$$= m\dot{\mathbf{v}}\cdot\mathbf{v} + \frac{1}{2}\frac{dm}{dt}(\mathbf{v}\cdot\mathbf{v}). \quad (23)$$

Comparison with equation (22) brings out the relation

$$\mathbf{R}\cdot\mathbf{v} = \left(m\dot{\mathbf{v}} + \frac{dm}{dt}\mathbf{v}\right)\cdot\mathbf{v} = m\dot{\mathbf{v}}\cdot\mathbf{v} + \frac{dm}{dt}\mathbf{v}\cdot\mathbf{v}$$

$$= \frac{dE}{dt} + \frac{1}{2}\frac{dm}{dt}\mathbf{v}\cdot\mathbf{v}. \quad (24)$$

And once more a variation from the previous model is impressed upon us; the power (P) is thrown out of equivalence with the activity or working-rate of the force (**R**), thus realizing the suggested contingency of section 30. The time-integral of the last equation assumes the form

$$\int_{t_1}^{t_2}(\mathbf{R}\cdot\mathbf{v})dt = [E]_{t_1}^{t_2} + \tfrac{1}{2}\int_{t_1}^{t_2}(dm\mathbf{v}\cdot\mathbf{v}), \quad (25)$$

and expresses on its face the conclusion that the total work of the force (**R**) for the interval is not accumulated in the change shown by the kinetic energy. What the form and the fate are of the energy summed in the last integral remains as a physical question for further study; it may, for instance, cease to be available, or it may be stored reversibly ready to appear again by transformation.

If instead of dealing with the resultant (**R**) we proceed by the standard resolution into tangential and normal parts, these are

$$\mathbf{R}_{(t)} = \frac{dm}{dt}\mathbf{v} + m\dot{\mathbf{v}}_{(t)}; \qquad \mathbf{R}_{(n)} = m\dot{\mathbf{v}}_{(n)}; \tag{26}$$

and if we should maintain that measure of force which is expressible as the product of mass and its acceleration, the inferences from the above equations would lead through the quotients of force by its acceleration to different estimates of the mass involved. From the first equation we obtain as a ratio of tensors

$$\frac{R_{(t)}}{\dot{v}_{(t)}} = \frac{dm}{dv}v + m, \qquad \text{since} \qquad \dot{v}_{(t)} = \frac{dv}{dt}; \tag{27}$$

and from the second equation

$$\frac{R_{(n)}}{\dot{v}_{(n)}} = m. \tag{28}$$

33. The last value agrees with our initial supposition, and is to that extent the true mass; and the value given by the first quotient in equation (27) has been distinguished as effective mass since the motion of a submerged body through a liquid suggested the term. We are aware how that idea has been borrowed and systematized in connection with the dynamics of electrons; and it is, therefore, of interest to verify that the difference between *longitudinal mass* and *transverse mass* originally introduced there, though now perhaps in course of abandonment, is quantitatively identifiable with the term (vdm/dv) according to the assumed relations for electrons of dependence of mass upon speed.

The effect when we are conscious of the whole situation must be to make evident how much turns upon attributing the entire force (**R**) to the mass (m), because a force diminished by the amount of the last term in equation (22) would reëstablish conformity with the type of equation (16) as

$$(\mathbf{R} - \Delta\mathbf{R}) = m\dot{\mathbf{v}}; \qquad \Delta\mathbf{R} = \frac{dm}{dt}\mathbf{v}. \tag{29}$$

And this is not mere mathematical ingenuity, for in the hydrodynamical conditions at least we know that the excess of effective mass over the weighed mass is only a disguised neglect of backward force upon the advancing body due to displacement of the liquid. So that while groping among phenomena that are less understood, our attention should keep equal hold upon both alternatives of statement until experimental analysis decides finally between them. It is in some degree a question of words whether all of the force (**R**) falls within a specified boundary.

34. The formal changes that have been pointed out, and their possible reconcilement with a larger group of facts through a second physical view, are important enough to justify this immediate effort to fix attention upon them. The path is beset with similar ambiguities whenever the details attendant upon transformations of the subtler forms of energy are sought. Therefore it is vital to pursue the thought of the section referred to, and to perceive with conviction even in this simplest example offered, how the bare assertion that a time rate for mass will be introduced for better embodiment of the data leaves the dynamics still impracticably vague for decision. We could not pass upon the physical validity and sufficiency of the force (**R**) assigned by equation (22) without fuller insight into suppositions. The instinctive control of the mathematics by repeated references to the physics is so well worth strengthening that we shall dwell upon one other side of the instance before us, though for suggestion only and not with any elaborate intention of exhausting it.

35. If a stream of water flows steadily in straight stream lines and with equal velocity everywhere, there is no loophole for acceleration, neither of an individual particle nor in passage systematically from one to another. Yet under an arbitrary agreement to include more and more water in the stipulated boundary the total momentum would gain an assigned time rate and the (**R**) of equation (22) a value

$$\mathbf{R} = \frac{dm}{dt}\mathbf{v}. \tag{30}$$

This is plainly illusory and void of dynamical meaning. We must cut off *change of mass by mere lapse of time;* this is one wording of the conclusion. But on the other hand conceive the mass (m) to grow continuously by picking up from rest differential accretions, somewhat as a raindrop may increase by condensation upon its surface, and equation (30) traces a physical process.

Investigation of this as a physical action confirms equation (30) quantitatively for a proper surface distribution of the elementary impacts, as force called for if the slowing of speed is to be compensated that would be consequent upon redistribution of the same total momentum through a continuously increasing mass. Thus much of force being allotted to keeping the velocity of the growing system constant, only the margin above this part would be registered in the acceleration. Moreover the way is then opened to interpret the last term in equation (25) by adapting specially the usual expression for kinetic energy converted at impact into other forms. Quoting, in a notation that will be understood at sight, we write that *loss* in the form

$$L = (1 - e^2)\frac{m_1 m_2}{2(m_1 + m_2)}(v_1 - v_2)^2. \tag{31}$$

Applying this to the conditions of inelastic central impact $(e = 0)$; with the ratio (m_1/m_2) negligible, as (dm/m) is; and when the relative speed $(v_1 - v_2)$ is (v); we find

$$L = \tfrac{1}{2}dmv^2. \tag{32}$$

And this wastage of kinetic energy finds due representation through the integral in question.

The essential condition, however, about (L) is a conversion of kinetic energy; and as remarked already that conversion might just as well be reversible. It is, therefore, again suggestive and perhaps even significant, that the sharing of energy between two forms indicated in the second member of equation (25) can be seen to correspond quantitatively with the partition of energy between the electric and the magnetic field of an electron as authoritatively calculated according to the assumed rate of change in its mass with speed. Of course this verifies or proves nothing, except the possibility in this direction as in others of constructing a mechanical process that is quantitatively adjusted to other and different processes where energy is converted.[1]

36. The six chosen quantities have been made definite by means of defining equations, which are truly designated as fundamental to the degree that the quantities involved possess that quality. With these identities we have been content to occupy ourselves mainly thus far, and confine discussion to phenomena observed or observable in a system of bodies, and to be described in terms of the masses, their radius-vectors, and two derivatives of the latter. With data of this type a range of inferences can be drawn, quantitatively determinate, too, up to a certain point, regarding the physical influences under which the system will furnish those data. Any assumed local distribution of mass, velocity and acceleration demands calculable aggregates of force, momentum and the rest, which the equations can be taken to specify. But nowhere along this line of thought is the further question mentioned, about how the influences shall be provided and brought to bear in producing what we see and measure, or what is visible and measurable in the system that is under observation. Not that the relations prove finally to be so one-sided as the sequence of our mathematics would suggest, according to which it happens that first mention is given to (**Q**, E, **H**), and they are made primary in the sense that the group (**R**, P, **M**) then follow by differentiation.

[1] See Note 11.

Yet the latter group would precede more naturally if the object were to reach the first group by integration; and this inverse order is revealed to be also a normal alternative. That procedure erects into data the physical influences like Force, Power and Force-moment to which the system is externally or internally subjected, and makes attack in the direction of predicting the response of the system in detail. The unconstrained tendency of this line of approach is then to set forth the supplementary idea that the accumulations of Momentum, Kinetic Energy and Moment of Momentum in the system of bodies are to be read as integrated consequences of the influences first specified.[1]

The formal change is inconsiderable, though the spirit of it guides three of our announced identities into full-fledged equations either of whose members is calculable in terms of the other. By usual title, these are the Equations of Motion, Work and Impulse that are an important part of dynamical equipment and that will next engage our attention. Since deciphering and listing the operative physical conditions comes now to the front, the weighing of arguments converges upon making the list of forces that is sought exhaustive, and upon weeding out illusory items from it. It must be apparent how that search and critical revision are bound up with inquiries like the suggestions of the previous section.

37. Dynamical analysis of results in its field has everywhere made tenable and corroborated the thesis that momentum and kinetic energy are traceable as fluxes. This is understood to imply that each local increase of those quantities will be found balanced against some other local decrease, either manifest in the quantity as such, or finally detectable under certain disguises of transformation. In application to a system of bodies, this means identifying a process of exchange dependent upon what is

[1] See Note 12.

in some sort external to it, and sometimes located to occur over the whole boundary or over limited areas of it, or sometimes recognized to permeate the whole volume or limited regions of it. Under the conditions that go with change in total mass by the passage of material through the stipulated boundary, the mass thus gained or lost may just carry its momentum and kinetic energy out or in, without any complicating interactions.

If, however, we exclude and put aside such processes of pure convection by confining ourselves to complete mass-constancy, there is evidence that changes in the total kinetic energy and the total momentum of a system of bodies are accompanied universally by exhibitions of force at the seat of the transfer. And this remains equally true whether a transformation between other recognizable forms and the mechanical quantities denoted by (**Q**) and (E) is taking place there or not. The possible exchanges between kinetic energy and other types, and the change-ratios corresponding to them are a commonplace of modern physics; as also we know how refined measurement has attested the forces upon bodies at transformations like that into light-energy. The settled anticipations in those respects have become even strong enough to look confidently upon occasional failure as only postponed success. The more recent proposal is to include momentum as well as kinetic energy within the scope of these ideas and concede for both alike a conversion into less directly sensible modifications, with force exerted upon bodies of the system or by them as a symptom of the transformation. And there seems to be no cogent reason why this should not hold its ground.

38. The quantitative formulation of these two transfers by flux in relation to what we shall call the transfer-forces temporarily and for the purpose of present emphasis because they are symptomatic of such action, presents to us the familiar equations of impulse and work which shall be first written, with the usual mass-constancy supposed, in the forms

$$\mathbf{Q} - \mathbf{Q}_0 = \Sigma \int_0^t d\mathbf{R}'dt \text{ (The Equation of Impulse);} \quad (33)$$

$$E - E_0 = \Sigma \int_0^{s'} d\mathbf{R}' \cdot ds' \text{ (The Equation of Work).} \quad (34)$$

They are intended to express total change from ($\mathbf{Q}_0$) to ($\mathbf{Q}$) during any time-interval (0, t), and total change from (E_0) to (E) during any simultaneous displacements (0, s′) at the points of application of the transfer-forces ($d\mathbf{R}'$). The integrations then cover the summation of effect over time or distance for each differential force ($d\mathbf{R}'$); and the symbol (Σ), though open to mathematical criticism as a crude notation, is doubtless sufficiently indicative of a purpose to include the aggregate of all such forces at every area and volume where the transfers may be proceeding. We must make also the necessary discrimination between the forces denoted by ($d\mathbf{R}'$) and those symbolized by ($d\mathbf{R}$) in equation (16), that are localized by association with elements of mass and not by participation in some transfer process, and that express themselves through the local accelerations manifested within the material of the system of bodies, while the forces ($d\mathbf{R}'$) can be determined wholly or to an important degree by data extraneous to the system.

It should be remarked next how one summation prescribed by the second member of equation (33) can be executed without further knowledge or specification, since the one time-interval applies in common to all force elements ($d\mathbf{R}'$) that are making simultaneous contributions toward the total change ($\mathbf{Q} - \mathbf{Q}_0$). Hence if the vector sum of these forces in whatever distribution they occur be written ($\mathbf{R}'$), the explained sense of this addition standing entirely in parallel with the comment attached to ($\mathbf{R}$) in equation (16), we see that

$$\mathbf{Q} - \mathbf{Q}_0 = \int_0^t \mathbf{R}'dt. \quad (35)$$

A corresponding general reduction of equation (34) would first require equal vector displacements (ds) at all points of application throughout the group of (d**R**′), a condition that need not be satisfied.

A second essential difference between the equations of impulse and work is that the former includes indifferently every force (d**R**′), in that some duration of its action is a universal characteristic. But in order that a force (d**R**′) may be effective in work, not only must there be displacement at the point where it acts upon the system, but that displacement must not be perpendicular to the line of the force. Either of these conditions may be at variance with the facts. It is a convenient usage to distinguish transfer-forces as *constraints* when they do no work; which signifies also when their work is negligible, of course.

39. Both (**R**′) and (**R**) are vector sums and have been exposed in their formation similarly to cancellation, but there is no presupposed relation of correspondence in detail between the two groups that would coördinate the occurrence and the extent of such spontaneous or automatic disappearances from the two final totals. If however we begin by confining comparison to those totals as such, that is yielded through the correlation of two statements which are now before us. Form the time-derivative of equation (35), replacing (**Q**) by its defined general equivalent from equation (I) and repeating its conditional derivative from equation (16). The consequence to be read is

$$\dot{\mathbf{Q}} = \mathbf{R} = \frac{d}{dt}\int_0^t \mathbf{R}'dt = \mathbf{R}'; \qquad (36)$$

and the relation between the extreme members of the equality is contingent only upon the validity of equation (35). This would carry the equality unconditioned otherwise of (**R**′) and (**R**) if ($\dot{\mathbf{Q}} \equiv \mathbf{R}$) can be introduced as a defining general equation. It gives latitude enough for the present line of thought to accept (**R**) as first quoted.

On its surface the last equation offers the meaning that the forces applied to the system under the rubric (**R**′) are competent to furnish exactly the total of force exhibited through the constituents of (**R**). And the same leading idea dictates the other verbal formula: The forces (d**R**) are an emergence of the group (d**R**′) after a transmission and a local redistribution. But neither reading is a truism, as the world has realized since d'Alembert first made the truth evident; for equation (36) does no more than convey one fruitful aspect of d'Alembert's principle which declares equality for the *impressed forces* (**R**′) and the *effective forces* (**R**), which names sanctioned by general usage we shall now adopt, and standardize the relation as the *equation of motion* under the form

$$\Sigma(d\mathbf{R}') = \Sigma \int_m \dot{\mathbf{v}} dm. \quad \text{[The Equation of Motion.]} \qquad (37)$$

In the first member the sign (Σ) recurs to the intention explained for equations (33, 34); and the particular basis of the second member has been made part of the record.

It is already clear that we have now come to deal with an equation by whose aid can be calculated either what total of impressed force is adequate to produce designated accelerations in given masses or what distributions of accelerations throughout a mass are compatible with a known group of impressed forces as their consequence. But the predicated equality is restricted to the totals and contains that element of indeterminateness which affects every resultant, in so far as it is an unchanging representative of many interchangeable sets of components. And in any properly guarded terms that are equivalent to the statement made above, the acknowledged deduction from the equality is in its chief aspect a conclusion about the acceleration at the center of mass of the system when (**R**′) is known, or a foreknowledge of what (**R**′) must be somehow built up if that center of mass is to be accelerated according to a known rule.

40. If there were complete physical independence among the masses of a system, or, in the current phrase, if there were no connections and constraints active between them to hamper mutually the freedom of their individual motions, impressed forces would make their effects felt only locally where they were brought to bear. And then for each such subdivision of the total mass as was thus affected equation (37) would apply, and an impulse equation would follow. Observe however that the question of minuteness in the subdivision enters, and that practically halt will be made with some undivided unit, assigning to it a common value of acceleration; so that the center of mass idea reappears in this shape ultimately, and duly proportioned to the scale of force-distribution symbolized by (d**R**′).

In actual fact there are found to be connections among the parts of a system of bodies, whose local influence deflects the acceleration from being purely the response to the local quota of (d**R**′). In other words, the masses of the system can exercise upon each other a group of forces internally, which must be regarded as superposed upon the impressed forces before the account of locally active force is to be held complete. To be sure this reduces to the now almost instinctive perception that external and internal are relative in use, and that an action may be impressed from outside upon a part which is exercised internally in respect to a larger whole. But like many other simple thoughts it was once announced for the first time.

Now certain forces being impressed, and with whatever internal connections interposed that the system is capable of exercising, the net outcome is an observable group of effective forces. It is therefore common sense to conclude that this net effect could be entirely nullified, in respect to the accelerations produced locally, by a second group of impressed forces applied also locally, and everywhere equal and opposite to the local value given by (d**R**). In virtue of equation (37) moreover it becomes apparent that the supposititious second group of impressed forces would always amount in their aggregate to $(-\mathbf{R}')$. Hence two auxiliary conclusions can be stated: First and negatively, that the superposed internal connections do not on the whole modify the original net sum $(\mathbf{R}')$; and the second is positive, to the effect that the office of internal connections in these relations is to transmit and make effective where they would otherwise not be felt in the system, the distribution of impressed forces $(d\mathbf{R}')$.

The internal connections can be described legitimately as themselves in equilibrium; they are the *lost forces* of d'Alembert. And the really applied group $(d\mathbf{R}')$ would be in equilibrium also with our second group of locally impressed forces. But this compensation is a supposition contrary to fact; the resultant $(\mathbf{R}')$ is *unbalanced force* to use the ordinary phrase. These details of interpretation are requisite exposition of the formally insignificant change that writes instead of equation (37)

$$\begin{aligned} \Sigma(d\mathbf{R}') - \Sigma \textstyle\int_m \dot{\mathbf{v}}\,dm &= 0; \\ \Sigma(d\mathbf{R}' \cdot \delta\mathbf{s}') - \Sigma \textstyle\int_m (\dot{\mathbf{v}}\,dm) \cdot \delta\mathbf{s} &= 0; \end{aligned} \qquad (38)$$

as a formulation of d'Alembert's principle. The second form involves the so-called *virtual velocities* $(\delta\mathbf{s}', \delta\mathbf{s})$, which term is fairly misleading; for these symbols designate any displacements consistent with preserving the internal connections intact, and capable of occurring simultaneously; one group at the *driving points* of $(d\mathbf{R}')$ and the other locally at each (dm). Obviously either form aims to express that fictitious equilibrium which is derivable from the real conditions. Because the second form is cast into terms of work, it seems to call for the remark that the foundation upon which all of this is reared lies nevertheless in the impulse equation, and the development might be called an expansion of consequences under Newton's third law; there is no vital bearing upon the actual energy relations definitively established by it. What remains to be said in the latter respect we shall next consider.

41. The first and familiar fact is that the kinetic energy of a system of bodies can be affected by interactions that are usually styled internal: quotable instances being gravitational attraction between sun and earth, and the effects of resilience upon distorted elastic bodies. Therefore some deliberate caution must be observed in delimiting the terms external and internal in relation to impressed forces, if equation (34) is to cover the total change in kinetic energy and yet make no dislocation from the impulse equation. It will be noticed that the critical instances are connected with transformations of energy; and of energy that one mode of speech would describe as internal to the system.[1] We can put force exercised upon a body by action of the ether-medium into the other category, since that medium is by explicit supposition external to our conception of body.

The case of gravitation is resolved by the consideration that the conversion of its *potential energy* into the kinetic form is attended with exercise of equal and opposite forces upon two bodies, according to inference from observation. If both bodies are included in the system, these forces cancel each other and do not disturb previous conclusions; and if one body is outside the system's boundary, its action appears among the ($d\mathbf{R}'$). A parallel statement can be drawn up for elastic deformations; but there is a remnant of combinations that are more obscure, like the transformations of molecular and atomic energies that can also affect kinetic energy, and that are by common usage attributed to the system as an internal endowment. Our ignorance of their more intimate nature however does not seem a barrier; we can still look upon every change in a system's kinetic energy as accompanied by impressed forces ($d\mathbf{R}'$), whether these are exerted in self-compensated pairs and removed thus from

[1] See Note 13.

influence upon the impulse equation, or whether there are unbalanced elements that affect the total momentum in addition to changing the kinetic energy. To this extent all impressed forces can be called external, though there may be hesitation about classing as external or internal the particular type of energy that is under transformation to or from the kinetic form. The corollary may be added, that so long as equal and opposite elements of force are also colinear, their moments for any origin are self-cancelling; otherwise they constitute couples.

With the attempt to formulate correct equations of motion, the difficulties of physical dynamics may be said to begin, when it is required to make the list of impressed forces what we have spoken of as *exhaustive and freed from illusions.* Outside the range of rather direct perceptions, we grapple with uncertainties under conditions of imperfect knowledge—with hypothetical forces, intangible energies, figurative masses. Dynamics that was ready to renounce criticism of provisional equations of motion would be over-sanguine. Conversions of energy into the one distinctively mechanical form that we call kinetic are perhaps closest to direct inquiry into attendant circumstances; and though it would be overcautious to construct on that base only, it seems probable that dissecting there first is the clew to larger success, and that equations (33, 34, 36) are landmarks on that road.

In practice, the bare statement of d'Alembert's principle as given by any one of the three forms indicated is supplemented with some record of the particular connections that overcomes the difficulty of specifying every individual local acceleration, and reduces the number of indispensable data within manageable limits. The forces of the connections are thus described indirectly through the *geometrical equations of condition;* and this method is more effective than the more direct one, because the magnitude of the constraining forces will in general depend upon the speeds, though the kinematical analysis of the linkages remains unaltered. It is this thought that introduced Lagrange's use of *indeterminate multipliers.*[1]

None of these devices though qualifies the character of d'Alembert's equality in asserting a quantitative equivalence between a net total of external agency (impressed forces) and the response to it on the part of a system of bodies, as expressed in the states of motion that the effective forces summarize. The physical thought attaching to the equation of motion will be clearer when cause and effect are kept apart, and will tend toward obscurity or confusion when a shuffling of terms from one member to the other, as a mathematical device or for other reasons, has impaired this desirable homogeneousness.

42. One large section of dynamics is devoted to working out its principles in their application to rigid solids. As these are specified, they carry to an extreme limit a scheme of interconnections among their constituent parts that provides an ideal of internal structure which knows no rupture nor even distortion, but which provides inexorably all necessary constraining connections. Like other such concepts its consideration yields results which are not only valuable in themselves, but which also furnish a point of departure for the introduction of conditions that approach their standards closely enough to be taken account of by means of small corrective terms. Beside repeating that frequent and useful relation of a concept to actual data, the study of *rigid dynamics* has some more special reasons to support it, of which one is discoverable in the trend of theoretical views about the constitution of all systems of bodies. The boldest analysis of molar and molecular and atomic units, as a substratum for the increasing number of energy-forms that we associate with them and give passage through them, has not broken away entirely from utilizing rigid solids of smaller scale

[1] See Note 14.

and their dynamics. This gives the prevailing tone in attacking the atomic nucleus and its atmosphere of electrons even, with only such mental reactions to modify the trust in the details of the reasoning as have a wholesome influence to maintain the flexibility that is scientific and make our dynamics more nearly universal in what it embraces.[1] In this sense the kinematical phase, through which so many of these matters evolve, remains uncompleted—or we may dub it empirical—until dynamics can serve it with reasoned argument.

In the second place, however, any rule of constancy is likely to have an advantage of particular kind over the multifarious rules of variation in correlation with which it is unique. This goes beyond the formal gain in abolishing some mathematical complications, though that, too, frees our minds to entertain the salient ideas with fuller concentration. Like our previous assumption of constancy in mass, this added supposition of permanent internal arrangement puts off particularizing among rules of change, and enables us to carry forward through instructive developments the task of bringing some general principles more nakedly to discussion. This grows cumbrous or impossible where conclusions are subject to many contingent decisions.

43. It bears rather closely upon these suggestions that we can make one good entry upon the particular inquiries about rigid solids by resuming and continuing the line of thought that paused at equation (20). In that section some glimpses were secured of a superposition by means of which a serviceable sketch can be drawn of a dynamical outline for certain systems of bodies. Or otherwise stated, the actual totals of the important quantities are grouped round the concept of a representative particle, leaving only specified remainders for further consideration. Let us now separate from such a system one body that we shall suppose rigid and having continuous mass-distribution, and deduce for

[1] See Note 15.

it, with increased finality of detail, the special consequences that seem valuable for our purpose. It is clear that the center of mass of this body will retain all the functions already assigned to the representative particle, and also that it must now in addition, because the body is rigid, fall into an unchanging configuration that makes constant in length all such vectors as ($\mathbf{r}'$) of equation (12). And it follows too from the conception of rigidity that the internal connections are excluded from net effect upon the sequences of conversion that change the body's kinetic energy. They are reduced in their final influence to the office of transmitting and distributing the consequences of conversions and constraints that have been effected otherwise than by any machinery of readjustments, named or unnamed, of internal arrangement. The intended meaning is not essentially varied, though it has been rendered less explicit perhaps, when it is said that the impressed forces can here only displace the body as a whole, or that the internal connections can do no work.

44. Now it is the elementary characteristic of translation that it does apply to the body as a whole and affect it uniformly throughout in all kinematical respects. Our next natural step, therefore, is to examine the remaining possibility that is consistent with the constant length of every ($\mathbf{r}'$), and that therefore restricts the locus of each mass-element to some sphere that is centered on the center of mass. If we accept for this type of motion as a whole the term *rotation*, there still remain some particulars to establish definitely; and of these the first will be the general value of the velocity denoted by ($\mathbf{u}$) in equation (9), for which one fitting name is the *local velocity relative to the center of mass.* It is evidently identical with the local velocity ($\mathbf{v}$) of each (dm) if ($\bar{\mathbf{v}}$) is zero, or if the center of mass is the origin of reference. With control of the value for ($\mathbf{u}$) we can ultimately take up the evaluation of the terms that contain ($\mathbf{u}$) or depend upon it, knowing in advance that these can appear in (E, **H**, P, **M**) but not in (**Q**, **R**).

45. In order to approach the matter conveniently let (C′) denote the center of mass, and locate orthogonal axes there that are *lines of the body:* that is, they move with the body and retain their positions in it. The unit-vectors of those axes shall be (**i**′, **j**′, **k**′) in the standard right-handed cycle. Then using the word temporarily in an untechnical sense, any rotation relative to (C′) will in general change all the angles that (**i**′, **j**′, **k**′) make with the reference-axes. Consider first differential changes of orientation ($\boldsymbol{\alpha}$, $\boldsymbol{\beta}$, $\boldsymbol{\delta}$) matching the order of the unit-vectors. Then ($\boldsymbol{\alpha}$) as an angle-vector is normal to the plane of the consecutive positions of (**i**′); similarly for ($\boldsymbol{\beta}$) and (**j**′), and for ($\boldsymbol{\delta}$) and (**k**′). The corresponding linear displacements on unit sphere around (C′) are given as products of perpendicular factors by

$$d\mathbf{i}' = \boldsymbol{\alpha} \times \mathbf{i}'; \qquad d\mathbf{j}' = \boldsymbol{\beta} \times \mathbf{j}'; \qquad d\mathbf{k}' = \boldsymbol{\delta} \times \mathbf{k}'. \tag{39}$$

The vector products are not affected, and hence these equalities are not disturbed, if we introduce three arbitrary elements of angular displacement; ($\boldsymbol{\lambda}'$) in the line of (**i**′) into the first, ($\boldsymbol{\mu}'$) in the line of (**j**′) into the second, and ($\boldsymbol{\nu}'$) in the line of (**k**′) into the third, writing

$$d\mathbf{i}' = (\boldsymbol{\alpha} + \boldsymbol{\lambda}') \times \mathbf{i}'; \qquad d\mathbf{j}' = (\boldsymbol{\beta} + \boldsymbol{\mu}') \times \mathbf{j}'; \qquad d\mathbf{k}' = (\boldsymbol{\delta} + \boldsymbol{\nu}') \times \mathbf{k}'. \tag{40}$$

But because the axis-set must remain orthogonal in the rigid body, the elements of angular displacement in the line of the third axis must always be equal for the two other axes at the same stage. This renders possible the adjustments of particular values that make equations (40) simultaneous:

$$\boldsymbol{\lambda} = \boldsymbol{\beta}_{(\mathrm{i}')} = \boldsymbol{\delta}_{(\mathrm{i}')}; \qquad \boldsymbol{\mu} = \boldsymbol{\alpha}_{(\mathrm{j}')} = \boldsymbol{\delta}_{(\mathrm{j}')}; \qquad \boldsymbol{\nu} = \boldsymbol{\alpha}_{(\mathrm{k}')} = \boldsymbol{\beta}_{(\mathrm{k}')}; \tag{41}$$

with the consequence that equations (40) are satisfied in the forms

$$\mathrm{d}\mathbf{i}' = \mathrm{d}\boldsymbol{\gamma} \times \mathbf{i}'; \qquad \mathrm{d}\mathbf{j}' = \mathrm{d}\boldsymbol{\gamma} \times \mathbf{j}'; \qquad \mathrm{d}\mathbf{k}' = \mathrm{d}\boldsymbol{\gamma} \times \mathbf{k}'; \qquad \mathrm{d}\boldsymbol{\gamma} \equiv \boldsymbol{\lambda} + \boldsymbol{\mu} + \boldsymbol{\nu}. \tag{42}$$

The occurrence of the vector ($\mathrm{d}\boldsymbol{\gamma}$) as a common factor in all three equations, combined with its determination by projections on axes arbitrarily chosen and with the fact that simultaneous linear displacements at points in the same radius-vector must be proportional to distances from (C′), shows that at each epoch and for every ($\mathbf{r}'$) of constant length,

$$\mathrm{d}\mathbf{r}' = \mathrm{d}\boldsymbol{\gamma} \times \mathbf{r}'; \qquad \dot{\mathbf{r}}' = \mathbf{u} = \boldsymbol{\omega} \times \mathbf{r}'; \qquad \boldsymbol{\omega} \equiv \dot{\boldsymbol{\gamma}}. \tag{43}$$

Here ($\boldsymbol{\omega}$) denotes the *rotation-vector* for either body or axis-set, of course, since they are supposed to turn together. It follows without further question that if a rigid solid moves so that all its radius-vectors ($\mathbf{r}$) measured from any reference-origin remain of constant length, the simultaneous velocities ($\mathbf{v}$) of all mass-elements conform to the relation

$$\mathbf{v} = \boldsymbol{\omega} \times \mathbf{r}. \tag{44}$$

Any such motion as a whole is described as a pure rotation with angular velocity ($\boldsymbol{\omega}$), for which vector the origin is conventionally the base-point.

46. The vector ($\boldsymbol{\omega}$) is usually termed the *angular velocity of the body* at the epoch, the phrase being made reasonable by the appearance of ($\boldsymbol{\omega}$) as a factor common to all radius-vectors in equations like (43) or (44). But both the procedure by which this angular velocity was determined and its appearance in a vector product show plainly that its resultant value is not effective to produce changes of direction in all radius-vectors.[1] This common factor has been seen to include three elements that become superfluous each for one axis, as not influencing

[1] See Note 16.

angular displacement of it, nor the corresponding linear displacement of points in it. The rotation-vector is thus open to interpretation as a maximum value, useful in giving through its projection upon the normal to any plane at its base-point the part effective to bring about a complete angular displacement occurring in that plane. If we identify ($\boldsymbol{\omega}$) with the line of a rotation-axis, permanent or instantaneous, these explanations are consistent with the elementary ideas of spin about the rotation-axis and linear velocity given by the product of rate of spin and distance from the axis.

47. The preceding identification of a rotation-vector connects its considerations with departures from configurations of ($\mathbf{i}'\mathbf{j}'\mathbf{k}'$) that are themselves subject to self-produced change, in so far as they move with the body; and this might conceivably modify the result. But if that loop-hole seems to exist it is closed when we detect the same vector ($d\boldsymbol{\gamma}$) in direct terms of its projections upon the reference-axes oriented by ($\mathbf{ijk}$) permanently. And it is, further, worth while to do that, because these projections are uniquely advantageous in preparing for algebraic additions to express any resultant angular displacement according to the relation

$$\boldsymbol{\gamma} = \int d\boldsymbol{\gamma} = \mathbf{i}\int d\gamma_{(i)} + \mathbf{j}\int d\gamma_{(j)} + \mathbf{k}\int d\gamma_{(k)}, \tag{45}$$

the tensors that are integrated being those of the projections of each ($d\boldsymbol{\gamma}$) upon the axes of ($\mathbf{i}$, $\mathbf{j}$, $\mathbf{k}$). The confirmation sought depends upon satisfying the relations,

$$\left.\begin{aligned} d\gamma_{(i)} &= \lambda\mathbf{i}'\cdot\mathbf{i} + \mu\mathbf{j}'\cdot\mathbf{i} + \nu\mathbf{k}'\cdot\mathbf{i} \\ d\gamma_{(j)} &= \lambda\mathbf{i}'\cdot\mathbf{j} + \mu\mathbf{j}'\cdot\mathbf{j} + \nu\mathbf{k}'\cdot\mathbf{j} \\ d\gamma_{(k)} &= \lambda\mathbf{i}'\cdot\mathbf{k} + \mu\mathbf{j}'\cdot\mathbf{k} + \nu\mathbf{k}'\cdot\mathbf{k}. \end{aligned}\right\} \tag{46}$$

Ordinary routine verifies that equations (46) fulfil identically the necessary conditions:

$$\left.\begin{aligned} d\mathbf{i}' &= d\boldsymbol{\gamma} \times \mathbf{i}' = \mathbf{i}(i'_{(k)}d\gamma_{(j)} - i'_{(j)}d\gamma_{(k)}) \\ &\qquad + \mathbf{j}(i'_{(i)}d\gamma_{(k)} - i'_{(k)}d\gamma_{(i)}) \\ &\qquad\qquad + \mathbf{k}(i'_{(j)}d\gamma_{(i)} - i'_{(i)}d\gamma_{(j)}). \\ d\mathbf{j}' &= d\boldsymbol{\gamma} \times \mathbf{j}' = \text{etc.} \\ d\mathbf{k}' &= d\boldsymbol{\gamma} \times \mathbf{k}' = \text{etc.} \end{aligned}\right\} \quad (47)$$

It is not without interest to notice in detail how algebraic cancellations now preserve the obligatory independence of ($\boldsymbol{\lambda}$) in the results for ($d\mathbf{i}'$); of ($\boldsymbol{\mu}$) in those for ($d\mathbf{j}'$); and of ($\boldsymbol{\nu}$) in those for ($d\mathbf{k}'$). This second development is more circuitous, because the permanently orthogonal condition, due to rigidity, pertains intimately to ($\mathbf{i}', \mathbf{j}', \mathbf{k}'$), the coincidence of results by both attacks being a special instance under a general theorem that will be proved subsequently (see section 85). The equal corroboration of equation (44) is a plain inference, and hence, wherever a rotation-vector covers the local velocities of a rigid body, or the body is in *pure rotation about a fixed point*, the summed projections are invariant:

$$\boldsymbol{\omega}_{(i)} + \boldsymbol{\omega}_{(j)} + \boldsymbol{\omega}_{(k)} = \boldsymbol{\omega}_{(i')} + \boldsymbol{\omega}_{(j)'} + \boldsymbol{\omega}_{(k')} = \boldsymbol{\omega}. \quad (48)$$

Substitute in equation (44), use the standard relation for common origin,

$$\mathbf{r} = \mathbf{x} + \mathbf{y} + \mathbf{z} = \mathbf{x}' + \mathbf{y}' + \mathbf{z}', \quad (49)$$

and omit products of colinear factors. This yields

$$\left.\begin{aligned} \mathbf{v} &= \boldsymbol{\omega}_{(i)} \times (\mathbf{y} + \mathbf{z}) + \boldsymbol{\omega}_{(j)} \times (\mathbf{z} + \mathbf{x}) + \boldsymbol{\omega}_{(k)} \times (\mathbf{x} + \mathbf{y}) \\ &= \boldsymbol{\omega}_{(i')} \times (\mathbf{y}' + \mathbf{z}') + \boldsymbol{\omega}_{(j')} \times (\mathbf{z}' + \mathbf{x}') + \boldsymbol{\omega}_{(k')} \times (\mathbf{x}' + \mathbf{y}'), \end{aligned}\right\} \quad (50)$$

and is the foundation for a standard rule: Linear velocities in a rotating rigid body are given correctly by superposing those due to separate partial rotations, either about the reference-axes or about the positions at the epoch of any three lines of the body intersecting orthogonally at the origin.

48. In the present connection however we are dealing with a rotation relative to (C′) as superposed upon the concept of a representative particle and supplementing the latter, with a proved equivalence of translation and rotation thus combined in replacing the most general group of velocities in our rigid body. On incorporating these recent restatements into equations (10) and (12), they take on the more special forms that we can now exhibit. Denote the last terms in the two equations by (E_R) and ($\mathbf{H}_R$), which we shall call briefly the kinetic energy and the moment of momentum relative to the center of mass. Then for the one body of continuous mass

$$\mathbf{H}_R = \int_m(\mathbf{r}' \times \mathbf{u}dm) = \int_m(\mathbf{r}' \times (\boldsymbol{\omega} \times \mathbf{r}')dm)$$
$$= \int_m(\boldsymbol{\omega}(\mathbf{r}'\cdot\mathbf{r}') - \mathbf{r}'(\boldsymbol{\omega}\cdot\mathbf{r}'))dm; \quad (51)$$

$$E_R = \tfrac{1}{2}\int_m\mathbf{u}\cdot\mathbf{u}dm = \tfrac{1}{2}\int_m(\boldsymbol{\omega} \times \mathbf{r}')\cdot(\boldsymbol{\omega} \times \mathbf{r}')dm$$
$$= \tfrac{1}{2}\int_m((\omega r')^2 - (\boldsymbol{\omega}\cdot\mathbf{r}')^2)dm = \tfrac{1}{2}(\boldsymbol{\omega}\cdot\mathbf{H}_R); \quad (52)$$

the final reduction of (E_R) being readily verifiable, when we remember that ($\boldsymbol{\omega}$) is common to all elements in these mass-summations.

49. Next we continue into equations (17) and (18) the same plan of partition between representative particle and supplementary term. Direct substitution there according to the relations previously used,

$$\mathbf{v} = \bar{\mathbf{v}} + \mathbf{u}; \qquad \mathbf{r} = \bar{\mathbf{r}} + \mathbf{r}'; \quad (53)$$

gives

$$P \equiv \frac{dE}{dt} = \bar{\mathbf{v}}\cdot\mathbf{R} + \int_m\mathbf{u}\cdot d\mathbf{R}; \quad (54)$$

$$\mathbf{M} \equiv \dot{\mathbf{H}} = (\bar{\mathbf{r}} \times \mathbf{R}) + \int_m(\mathbf{r}' \times d\mathbf{R}). \quad (55)$$

We may remind ourselves that the first terms in the final members of both these equations are in harmony with the time-derivatives of corresponding terms in equations (10) and (12) if we bear in mind equation (20); and they show how the particle can be relied upon still to present these contributions to power and to force-moment as based upon its artificial translation with the center of mass. Denote the additional power and force-moment by (P_R) and ($\mathbf{M}_R$); then from equations (54, 55),

$$P_R = \int_m(\boldsymbol{\omega} \times \mathbf{r}')\cdot d\mathbf{R} = \boldsymbol{\omega}\cdot\int_m(\mathbf{r}' \times d\mathbf{R}) = \boldsymbol{\omega}\cdot\mathbf{M}_R; \qquad (56)$$

$$\mathbf{M}_R \equiv \int_m(\mathbf{r}' \times d\mathbf{R}). \qquad (57)$$

We shall compare these statements with the consequences of equations (51, 52), which give for their derivatives

$$\frac{d}{dt}(E_R) = \tfrac{1}{2}(\dot{\boldsymbol{\omega}}\cdot\mathbf{H}_R + \boldsymbol{\omega}\cdot\dot{\mathbf{H}}_R); \qquad (58)$$

$$\dot{\mathbf{H}}_R = \frac{d}{dt}\int_m(\mathbf{r}' \times \mathbf{u}dm) = \int_m(\mathbf{r}' \times \dot{\mathbf{u}}dm); \qquad (59)$$

because ($\mathbf{u}$) and ($\dot{\mathbf{r}}'$) are identical. Further, since differentiation of equation (9) shows

$$\dot{\mathbf{v}} = \dot{\bar{\mathbf{v}}} + \dot{\mathbf{u}}, \qquad (60)$$

a natural name for the last term is the *local acceleration relative to the center of mass*, which would indicate also a local force-element ($\dot{\mathbf{u}}dm$) differing from ($\dot{\mathbf{v}}dm$) that is ($d\mathbf{R}$) and thereby breaking the equality of ($\dot{\mathbf{H}}_R$) and ($\mathbf{M}_R$). But since

$$\int_m\mathbf{r}'dm = 0, \qquad (\int_m\mathbf{r}'dm) \times \dot{\bar{\mathbf{v}}} = \int_m(\mathbf{r}' \times \dot{\bar{\mathbf{v}}}dm) = 0; \qquad (61)$$

and this term can be added without error to equation (59), giving

$$\dot{\mathbf{H}}_R = \int_m(\mathbf{r}' \times (\dot{\bar{\mathbf{v}}} + \dot{\mathbf{u}}))dm = \int_m(\mathbf{r}' \times d\mathbf{R}) = \mathbf{M}_R. \qquad (62)$$

Evidently the value in equation (61) could reversely be subtracted without error from equation (57). The interchangeableness of these forms should not be lost sight of.

50. A similar concordance of equations (56, 58), though it is not superficially evident, follows at once on showing a right to add the third member in the equality

$$\boldsymbol{\omega}\cdot\mathbf{M}_R = \boldsymbol{\omega}\cdot\dot{\mathbf{H}}_R = \dot{\boldsymbol{\omega}}\cdot\mathbf{H}_R, \tag{63}$$

whose first and second members are now known to be equal. The required proof is got by differentiating equation (51), where we find

$$\dot{\mathbf{H}}_R = \int_m\{\dot{\boldsymbol{\omega}}(\mathbf{r}'\cdot\mathbf{r}') - \mathbf{u}(\boldsymbol{\omega}\cdot\mathbf{r}') - \mathbf{r}'(\dot{\boldsymbol{\omega}}\cdot\mathbf{r}')\}dm, \tag{64}$$

whose scalar product with ($\boldsymbol{\omega}$) is, omitting everywhere scalar products of perpendicular factors,

$$\begin{aligned}\boldsymbol{\omega}\cdot\dot{\mathbf{H}}_R &= \int_m\{(\boldsymbol{\omega}\cdot\dot{\boldsymbol{\omega}})(\mathbf{r}'\cdot\mathbf{r}') - (\boldsymbol{\omega}\cdot\mathbf{r}')(\dot{\boldsymbol{\omega}}\cdot\mathbf{r}'))\}dm \\ &= \int_m\dot{\boldsymbol{\omega}}\cdot(\boldsymbol{\omega}(\mathbf{r}'\cdot\mathbf{r}') - \mathbf{r}'(\boldsymbol{\omega}\cdot\mathbf{r}'))dm = \dot{\boldsymbol{\omega}}\cdot\mathbf{H}_R.\end{aligned} \tag{65}$$

The vector ($\dot{\boldsymbol{\omega}}$) which is the time-derivative of the rotation-vector ($\boldsymbol{\omega}$) is named the vector of *angular acceleration.* Of course it provides for both changes of direction (or of axis) in the rotation, and for changes in its magnitude (or spin); and ($\dot{\boldsymbol{\omega}}$) must be of common application at any epoch to all mass-elements, because that is true for ($\boldsymbol{\omega}$).

51. With the support of equations (51, 52, 56, 58), we have given consideration to all four quantities that need specifying, for the rotation that is the remainder over and above the fictitiously segregated translation, since the representative particle as it has been determined engages the totals of force and momentum. And having brought the discussion to this point, in terms connected with the effective forces whose resultant is (**R**), it remains to make that transition to impressed forces with equal resultant (**R**′), which we have learned to associate with d'Alembert's name. Under the conditions explained for rigid bodies, certain sources of impressed force are not to be permitted, but the total work done must appear in the energies of translation and rotation. Let us then next summarize how matters stand with the six dynamical quantities, in the two groups that we have recognized.

I. Translation:

1. Force ($\mathbf{R}' = \mathbf{R}$) at ($\bar{r}$).
2. Momentum ($\mathbf{Q} = m\bar{v}$) at ($\bar{r}$).
3. Energy ($E_T = \frac{1}{2}m\bar{v}^2$).
4. Moment of Momentum ($\mathbf{H}_T = \bar{r} \times m\bar{v}$); consistent with (2).
5. Power ($P_T = \mathbf{R}' \cdot \bar{v} = (d/dt)(E_T)$); consistent with (1) and (3).
6. Force-moment ($\mathbf{M}_T = \bar{r} \times \mathbf{R}' = \dot{\mathbf{H}}_T$); consistent with (1) and (4).

II. Rotation:

1. Force = 0; consistent with couples expressing self-compensating elements in ($\mathbf{R}'$).
2. Momentum = 0 always; consistent with impulse of zero force.
3. Energy ($E_R = \frac{1}{2}\omega \cdot \mathbf{H}_R$).
4. Moment of Momentum ($\mathbf{H}_R$); consistent with zero momentum.
5. Power ($P_R = \omega \cdot \mathbf{M}_R = (d/dt)(E_R)$); consistent with (1) and (3).
6. Force-moment ($\mathbf{M}_R = \dot{\mathbf{H}}_R$); consistent with (1), (3) and (5).

52. The review of these details impresses the fact that the above conventional separation accomplishes complete independence for two such constituents of the actual data, in the sense that the course of events can be duly expressed for each group, with indifference to the presence or absence of the other, by a self-contained use of the general dynamical scheme. This secures the full simplicity attendant on pure superposition, by shrewdly exploiting center of mass for its average properties, and kinetic energy with moment of momentum for their salvage of what the mean values sacrifice, utilizing also a form of Poinsot's allowance through a couple for off-center action of a force. The idea is successful, besides, in concentrating into the rotation elements where the form and the mass-distribution of the body complicate the data with differences; and this frees the translation for giving expression to broad traits of similarity.

The rudiments of the steps now taken are perceivable in equations (10) and (12), where it is plain that an internal energy like (E_R) could belong to radial pulsations of mass-elements about (C′), either alone or added to spin as a whole; but development is checked until (**u**) is particularized in its value and distribution. It is plain, however, that adaptation to many combinations is feasible, whose general feature is non-appearance in translational energy of full equivalent for the total work done. Failing definite knowledge that forbids, a rotation can be devised as one possible means of absorbing a quota of kinetic energy, and as one guide to conjecture among the facts of an observed diversion of energy from a translation. It is scarcely necessary to insist that the equivalence of any such devices is restricted to those particulars according to which their lines were laid down; the particle plus a rotation is an equivalent for the general motion of a rigid body only in the six respects enumerated.[1]

53. At equation (44) the idea was introduced that pure rotation of a rigid body about a reference-origin, instead of the center of mass, is describable in corresponding terms on substituting (**r**) for (**r**′) and (**v**) for (**u**). The intrinsic difference lies in the necessity that a reference-origin is a fixed point, whereas the possible velocity of the center of mass runs like a thread through all our recent discussion. Let us realize that the main results now added can be similarly extended, and put down as applicable to pure rotation about the reference-origin these parallels specifically to equations (51, 52, 56, 62, 65):

[1] See Note 17.

$$\left.\begin{aligned} \mathbf{H} &= \textstyle\int_m(\omega(\mathbf{r}\cdot\mathbf{r}) - \mathbf{r}(\omega\cdot\mathbf{r}))dm; \\ \mathrm{E} &= \tfrac{1}{2}(\omega\cdot\mathbf{H}); \\ \mathrm{P} &= \omega\cdot\mathbf{M}; \\ \dot{\mathbf{H}} &= \mathbf{M}; \\ \dot{\omega}\cdot\mathbf{H} &= \omega\cdot\dot{\mathbf{H}} = \omega\cdot\mathbf{M}. \end{aligned}\right\} \text{Total quantities for pure rotation.} \quad \begin{aligned} &(66)\\ &(67)\\ &(68)\\ &(69)\\ &(70) \end{aligned}$$

Since in this case supposed, the center of mass need not coincide with the origin, the alternative choices will be open to treat the body as exhibiting rotation alone, or as affected with translation and with a rotation besides. But translation cannot bring in change of direction for lines of the body, hence both views of the rotation must agree in their rotation-vectors permanently. And because the center of mass cannot change its position relative to its rigid body, a relation distinctive of pure rotation must be

$$\bar{\mathbf{v}} = \omega \times \bar{\mathbf{r}}. \tag{71}$$

The comparative directness and convenience of the two methods will be decided according to circumstances. One method excludes from (**M**) any forces really acting through the origin; the other can omit from ($\mathbf{M}_R$) any forces acting through (C′).

54. We proceed with the requisite analysis of rotation, by examining the specialized values of local accelerations and some consequences of them, conscious always in the light of what has just been said, that the conclusions will be available for twofold use. One is more important, doubtless, because more inclusive in application to the most general type of motion of which a rigid body is capable; but the second has weight, too, in attacking the conditions of pure rotation that are made prominent, for instance, in common forms of the gyroscope.

The local acceleration of a pure rotation given by differentiating equation (44) is

$$\dot{\mathbf{v}} = (\dot{\omega} \times \mathbf{r}) + (\omega \times \mathbf{v}). \tag{72}$$

Let us make this form our text and starting-point, remembering that in the other circumstances it is to be recast into

$$\dot{\mathbf{u}} = (\dot{\boldsymbol{\omega}} \times \mathbf{r}') + (\boldsymbol{\omega} \times \mathbf{u}), \tag{73}$$

with continuations where $(\mathbf{r}')$ replaces $(\mathbf{r})$ everywhere and $(\mathbf{u})$ replaces $(\mathbf{v})$, while $(\dot{\mathbf{u}})$ is read the *local acceleration of the rotation* and is the excess of $(\dot{\mathbf{v}})$ over $(\dot{\bar{\mathbf{v}}})$. The vector $(\dot{\boldsymbol{\omega}})$ gives the velocity of the extremity of $(\boldsymbol{\omega})$, of course; and its base-point will be taken conventionally at the origin with which our idea of rotation is associated. Then the process modifying $(\boldsymbol{\omega})$ by $(\dot{\boldsymbol{\omega}})$ is one of continuous parallelogram composition for intersecting vectors, though equivalent indeed to addition in a triangle.

The vector sum in equation (72) deserves close attention, because though the two types of its terms are on one count an incident of the algebra, it happens that they conform remarkably, first to the kinematical elements, and later to a certain plane of cleavage in the dynamics. The form of the second term connects it conclusively with change of direction only for its velocity; and the first term enters and vanishes with angular acceleration. If $(\boldsymbol{\omega})$ retains direction $(\dot{\boldsymbol{\omega}})$ must be colinear with it; and then first inspection can identify the terms with the tangential and the normal acceleration respectively of the local (dm) in its circle perpendicular to $(\boldsymbol{\omega})$. But the complete separation of changes in magnitude and in direction for $(\mathbf{v})$ that then exists should not be assumed more generally; it is always true, however, that the first term in the acceleration bears the same relation to the *axis of angular acceleration* $(\dot{\boldsymbol{\omega}})$ that the corresponding velocity $(\mathbf{v})$ does to the *axis of rotation* $(\boldsymbol{\omega})$.

55. Multiplying equation (72) by (dm) yields the effective force-element, which, because it is exhibited locally, must have a moment to be found by taking that force in vector product with its $(\mathbf{r})$. The total moment then demanded by the localized forces must, as we have seen, be furnished by the impressed forces; and this amount is expressed by the integral

$$\mathbf{M} = \int_m [\mathbf{r} \times ((\dot{\boldsymbol{\omega}} \times \mathbf{r}) + (\boldsymbol{\omega} \times \mathbf{v}))\mathrm{dm}]. \tag{74}$$

Denote the two main constituents of this moment by ($\mathbf{M}'$) and ($\mathbf{M}''$); and let us take up the second part for examination. Expand the triple vector product, omit the scalar product of perpendicular factors, and finally write for ($\mathbf{v}$) its known value. This shows

$$\mathbf{M}'' = -\int_m \mathbf{v}(\boldsymbol{\omega}\cdot\mathbf{r})dm = -\int_m(\boldsymbol{\omega}\times\mathbf{r})(\boldsymbol{\omega}\cdot\mathbf{r})dm. \qquad (75)$$

Next form for comparison the product

$$\boldsymbol{\omega}\times\mathbf{H} = \int_m(\boldsymbol{\omega}\times[\boldsymbol{\omega}(\mathbf{r}\cdot\mathbf{r}) - \mathbf{r}(\boldsymbol{\omega}\cdot\mathbf{r})]dm)$$
$$= -\int_m(\boldsymbol{\omega}\times\mathbf{r})(\boldsymbol{\omega}\cdot\mathbf{r})dm, \qquad (76)$$

and we see that the extreme members are identical. Hence we conclude that the office of thus much of the force-moment is to produce a change of direction in the vector of total moment of momentum so regulated that the latter would move with the body or retain its position in the body. This is a simple corollary of the interpretation of ($\boldsymbol{\omega}$) according to section (47). If ($\boldsymbol{\omega}$) and ($\mathbf{H}$) were in every case colinear, their vector product at the value zero would become formal and meaningless. But it appears plainly in equation (66), first that ($\mathbf{H}$) may be thrown out of line with ($\boldsymbol{\omega}$) by the term

$$-\int_m \mathbf{r}(\boldsymbol{\omega}\cdot\mathbf{r})dm,$$

which does not in fact generally vanish nor become colinear with ($\boldsymbol{\omega}$); and secondly, that ($\mathbf{H}$) and ($\boldsymbol{\omega}$) cannot become perpendicular by compensations within the first term, because every product ($\mathbf{r}\cdot\mathbf{r}$) is essentially positive. That they never are perpendicular we shall conclude presently (see section 58); the general obliquity of the rotation-vector and the moment of momentum vector is one characteristic in rotation, and is operative to cause effects to which there is no parallel where a kinematical vector and its dynamical associate are colinear, like momentum and its velocity. If angular acceleration is absent, every element in ($\mathbf{M}'$)

is zero, but ($\mathbf{M}''$) is not affected, since it depends upon the (ω) of the epoch, and not upon the past or future history of (ω). If a rigid body is spinning steadily about a fixed axis even, ($\mathbf{M}''$) is called for, as a *directive moment,* whenever (ω) and ($\mathbf{H}$) diverge. For the case of rotation about the center of mass, ($\mathbf{M}_R''$) will be furnished by a couple. These moments are recognizable as the *centrifugal couple* of the older fashion in speech. Like forces normal to a path, they disappear from the power equation by a condition of perpendicularity, as is visible from equation (68), when we have noticed through equations (75, 76) that ($\mathbf{M}''$) is perpendicular to (ω).

56. What has been determined about ($\mathbf{M}''$) presents it in such relation to the (ω) of the epoch that an impressed total force-moment of that value is adjusted exactly to continuance of constancy in the rotation-vector (ω); the zero value of power and the consequent constancy of (E) being an evident concomitant of that as primary condition. It is further acceptable on commonsense grounds that ($\mathbf{H}$) whose divergence from (ω) is fixed by the mass-distribution when (ω) is constant, as the form of equation (66) proves, must then accompany that mass-distribution through its changes in azimuth round the rotation-axis, so as to describe a right circular cone and keep up with any originally coincident radius-vector of the body. And the shrinking of such a cone into its axis provides for the singular case of non-divergence, with no ($\mathbf{M}''$) required for adjustment.

With the above details in hand, the part ($\mathbf{M}'$) of the force-moment appears in the light of a disturber of adjustment, and that opens for it an indefinite range of possibilities or puts away the expectation of particular conclusions, except two: that it must supply, first, all power and all changes in magnitude of ($\mathbf{H}$), and secondly, any change of direction that displaces ($\mathbf{H}$) relatively to the body.

57. At this point the chance offers for a pertinent remark about all equations like (74) in their type. They exhibit an impressed physical agency (here of (**M**)) in terms that compare it for excess or defect with an adjustment that is not compensation as equilibrium is, but calls for positive action (such as (**M**″) here exerts). It is an ambiguity inseparable from the algebra, especially where the total available is numerically less than the critical value, that an adjustment disturbed is indistinguishable from one not secured. In other words we can be sure only that (**M**′) and (**M**″) are mathematically represented in (**M**), when the latter has been assigned arbitrarily; using again the present instance, we know nothing of (**M**′) and (**M**″) separately as active agencies. Neither of the forms

$$\mathbf{M} = \mathbf{M}''; \qquad \mathbf{M} - \mathbf{M}'' = 0; \tag{77}$$

indicates equilibrium, but both express a fulfilled adjustment, much as equation (36) was read. Both of the forms

$$\mathbf{M} = 0; \qquad \mathbf{M}' + \mathbf{M}'' = 0; \tag{78}$$

apply the condition of equilibrium to (**H**) in the sense of making it a constant vector. In these circumstances an angular acceleration that underlies (**M**′) will appear in the equations unless (**M**′) and (**M**″) are zero separately, which can be true only specially; and there is some trace of mathematical suggestion that this angular acceleration arises by give-and-take between (**M**′) and (**M**″) that diverts the latter from its original office of keeping (ω) constant.

Doubtless that instinctive view, if it exists, receives some support from knowledge of other conditions in which an active assignable force-moment is indispensable to the appearance of angular acceleration; and that is the root of the inclination to see paradox in the phenomena that realize the conditions of equation (78). But in consequence of the divergence already spoken of, if the (**H**) vector preserves its direction in the reference-frame while the body is in rotation, the vector (**ω**), oblique to it, will not be constant also, and accordingly there will be angular acceleration. This occurs spontaneously we might say, (**M**) being zero, in the absence of control that would be effective to keep (**ω**) constant and shift the burden of change upon (**H**). It makes the reasons for the apparently abnormal results more obscure, that the kinematical aspects depending upon (**ω**) and ($\dot{\boldsymbol{\omega}}$) are often patently visible, whereas the dynamical elements that really dominate are hidden from view.[1]

58. While we are laying emphasis upon the general separation of directions for (**ω**) and (**H**), it is proper to be aware how this works out only for the body as a whole through the mass-summation of (d**H**) and the introduction of the common rotation-vector, and does not appear in the local elements, that it is the object of that plan and its advantage to handle in one group and not individually. It was observed already in equation (2) which had not yet been narrowed to rotation, that for each (dm) its (d**H**) and its ($\dot{\boldsymbol{\gamma}}$) are coincident vectors, the latter lying in the normal to the plane (**r**, d**s**) and being attributed to the local (**r**) as its particular angular velocity. This lesson can now be repeated from equation (51) or (66), if we denote by ($\boldsymbol{\omega}_1$, $\mathbf{r}_1$, $\boldsymbol{\gamma}_1$) the unit-vectors of (**ω**) and of (**r**), and of the perpendicular to (**r**) in the plane (**ω**, **r**), noticing that for instance equation (66) can be written, if (α) is the angle (**ω**, **r**),

$$d\mathbf{H} = (\boldsymbol{\omega}_1(\omega r^2) - \mathbf{r}_1(\omega r^2 \cos\alpha))dm$$
$$= \boldsymbol{\gamma}_1(\omega r^2 \sin\alpha)dm = \dot{\boldsymbol{\gamma}}(r^2 dm). \quad (79)$$

It is instructive to see, next, how the body as a whole retains for its total moment of momentum in relation to its rotation-vector the same type as equation (79) shows; and this can be done by assembling the projections of every (d**H**) upon the direction of (**ω**). The result to be recorded for use is

[1] See Note 18.

$$\mathbf{H}_{(\omega)} = \int_m \omega_1(\omega r^2 - (\omega_1 \cdot \mathbf{r})(\omega \cdot \mathbf{r}))dm = \omega I_{(\omega)}, \quad (80)$$

expressed as we find, also as the product of an angular velocity and a moment of inertia about its axis, but both these factors now refer to the whole body, and this form excludes perpendicularity of (ω) and ($\mathbf{H}$).

Because ($\mathbf{H}$) is a sum into which the differently weighted elements ($\dot{\gamma}$) enter, and the weighting depends upon what happens to be the mass-distribution, the final result cannot be forced completely into any one mould, beyond the point here established; only we know that the rest of ($\mathbf{H}$) must be in the plane perpendicular to (ω). Therefore according to equation (67) we learn that

$$E = \tfrac{1}{2}(\omega \cdot \mathbf{H}) = \frac{\omega^2}{2} I_{(\omega)}, \quad (81)$$

which may also be inferred directly from equation (52), by a slightly varied reduction of the last member but one. Let us use the occasion to renew the reminder that the rotation relative to (C′) involves only a transfer to its notation of the details here attached to the other case.

59. A similar trend can be marked in the other partners ($\dot{\omega}$) and ($\mathbf{M}'$) which bring kinematics and dynamics into connection: an elementary type of expression which appears differentially then persists in application to the body as a whole, but with a supplement governed by the particular mass-distribution that produces obliquity of ($\mathbf{M}'$) and ($\dot{\omega}$). For the local element ($d\mathbf{M}'$) equation (74) leads by expansion to

$$d\mathbf{M}' = (\dot{\omega}(\mathbf{r} \cdot \mathbf{r}) - \mathbf{r}(\dot{\omega} \cdot \mathbf{r}))dm, \quad (82)$$

which it will be noted reproduces equation (66), except that ($\dot{\omega}$) has replaced (ω) throughout. Consequently equation (79) can be paralleled in the form

$$d\mathbf{M}' = (\dot{\omega}_1(\dot{\omega} r^2) - \mathbf{r}_1(\dot{\omega} r^2 \cos\beta))dm$$
$$= \mathbf{p}_1(\dot{\omega} r^2 \sin\beta)dm = (\dot{\omega}\sin\beta)\mathbf{p}_1(r^2 dm). \quad (83)$$

But $(\dot{\boldsymbol{\omega}}_1, \mathbf{r}_1, \mathbf{p}_1)$ are now unit-vectors for $(\dot{\boldsymbol{\omega}})$, $(\mathbf{r})$ and the perpendicular to $(\mathbf{r})$ in the plane $(\dot{\boldsymbol{\omega}}, \mathbf{r})$, and (β) denotes the angle $(\dot{\boldsymbol{\omega}}, \mathbf{r})$. It is plain that $(\dot{\omega} \sin \beta)\mathbf{p}_1$ is for each $(\mathbf{r})$ the effective part of $(\dot{\boldsymbol{\omega}})$, as $(\omega \sin \alpha)\boldsymbol{\gamma}_1$ is the locally effective projection of $(\boldsymbol{\omega})$, and that (r^2dm) is a moment of inertia for the axis $(\mathbf{p}_1)$. Thus the type is set for the corresponding expression in terms devised to apply to the body; and in fact we find

$$\mathbf{M}'_{(\dot{\omega})} = \int_m \dot{\boldsymbol{\omega}}(r^2 - (\dot{\boldsymbol{\omega}}_1 \cdot r)^2)dm = \dot{\boldsymbol{\omega}} I_{(\dot{\omega})}, \tag{84}$$

whose form excludes perpendicularity likewise for $(\dot{\boldsymbol{\omega}})$ and $(\mathbf{M}')$.

60. It can be conceded as one legitimate purpose of equations (80), (81) and (84) to extract from the more general treatment of rotation what residue of correspondence remains with those simpler forms that are met in uniplanar dynamics. Looking in that direction, the main difference can be localized in the addition of an independent axis of $(\dot{\boldsymbol{\omega}})$ to stand alongside the previous axis of $(\boldsymbol{\omega})$. But the greater enlightenment in the discussion comes from the insistence upon putting foremost the powerfully direct analysis, by means of the dynamical vectors $(\mathbf{H})$ and $(\mathbf{M})$ and their connections. This tends to make the kinematical vectors, and especially $(\dot{\boldsymbol{\omega}})$, rather subsidiary until restrictions upon the problem restore to them more nearly equal weight.

61. If we start again from equation (66) and enter upon the semi-cartesian expansion for the vector $(\mathbf{H})$ the first results found are

$$\left.\begin{aligned} \mathbf{H}_{(x)} &= \boldsymbol{\omega}_{(x)} \int_m (\mathbf{r} \cdot \mathbf{r})dm - \int_m \mathbf{x}(\boldsymbol{\omega} \cdot \mathbf{r})dm; \\ \mathbf{H}_{(y)} &= \boldsymbol{\omega}_{(y)} \int_m (\mathbf{r} \cdot \mathbf{r})dm - \int_m \mathbf{y}(\boldsymbol{\omega} \cdot \mathbf{r})dm; \\ \mathbf{H}_{(z)} &= \boldsymbol{\omega}_{(z)} \int_m (\mathbf{r} \cdot \mathbf{r})dm - \int_m \mathbf{z}(\boldsymbol{\omega} \cdot \mathbf{r})dm. \end{aligned}\right\} \tag{85}$$

These continue to assume pure rotation round the origin, the body being in a general orientation relative to the reference-frame (XYZ). Retaining one value of $(\boldsymbol{\omega})$ given in relation to (XYZ), the last terms in the second members are seen to depend upon the body's orientation, but the first terms are invariant for all such orientations. By a definite choice of orientation the last terms can always be remarkably simplified, and what are known as the *principal axes of inertia* for the origin will then coincide with the axes (XYZ). We presuppose the proof that there are never fewer than three orthogonal principal axes at every point that is in rigid configuration with a rigid body, and ordinary acquaintance with properties of the *ellipsoid of inertia* or momental ellipsoid; this material is standard and accessible.

In all three equations expand $(\boldsymbol{\omega}\cdot\mathbf{r})$ and reduce to the forms

$$\left.\begin{aligned} \mathbf{H}_{(x)} &= \mathbf{i}\{\omega_{(x)}I_{(x)} - \omega_{(y)}\int_m xydm - \omega_{(z)}\int_m zxdm\}; \\ \mathbf{H}_{(y)} &= \mathbf{j}\{\omega_{(y)}I_{(y)} - \omega_{(z)}\int_m yzdm - \omega_{(x)}\int_m xydm\}; \\ \mathbf{H}_{(z)} &= \mathbf{k}\{\omega_{(z)}I_{(z)} - \omega_{(x)}\int_m zxdm - \omega_{(y)}\int_m yzdm\}. \end{aligned}\right\} \quad (86)$$

The property of principal axes determines the disappearance of six integrals at the orientation where those lines of the body coincide with (XYZ). Supposing that coincidence, therefore, it becomes true that

$$\mathbf{H} = \boldsymbol{\omega}_{(x)}I_{(x)} + \boldsymbol{\omega}_{(y)}I_{(y)} + \boldsymbol{\omega}_{(z)}I_{(z)}. \quad \text{[Principal axes.]} \quad (87)$$

But ($\mathbf{H}$) can be represented invariantly by an indefinite number of groups of orthogonal projections, and for one group, which can be chosen at every epoch and for every ($\boldsymbol{\omega}$), the coincidences that simplify equation (87) will occur instantaneously. How and on what terms the advantage of the simplification can be made permanently available is a question to be taken up hereafter (see section 118); but some useful decisions follow immediately here.

62. And first, the possible extent is made evident of the cancellation ensuing through the difference between the two contributions to the second member of equation (66). It is indicated by the present remainder, in which all the terms are essentially positive, if we take the vector factors absolutely. Secondly, if we turn to kinetic energy, the aid given by adopting principal axes, there too, is apparent in reducing the number of terms in the expression. For whereas the expansion of equation (67) on the basis of equation (86) will yield nine terms that do not coalesce into fewer than six, the reduction of these to three is a consequence of equation (87), from which follows

$$E = \tfrac{1}{2}[(\omega_{(x)})^2 I_{(x)} + (\omega_{(y)})^2 I_{(y)} + (\omega_{(z)})^2 I_{(z)}]. \qquad \text{[Principal axes.]} \quad (88)$$

This again by deleting subtractive terms has regained parallelism with the case of translation and three orthogonal components of velocity except for the difference, irreducible in the general expression, between the uniform mass-factor (m) and the individual inertia-coefficients like ($I_{(x)}$).

63. In the third place, that similarity in type between equation (66) and equation (82) which has been relied upon before to abbreviate details can be employed again. Like equations (86) for (**H**) we can write for (**M**′)

$$\left.\begin{aligned} \mathbf{M}'_{(x)} &= \mathbf{i}\{\dot\omega_{(x)} I_{(x)} - \dot\omega_{(y)}\textstyle\int_m xydm - \dot\omega_{(z)}\int_m zxdm\}; \\ \mathbf{M}'_{(y)} &= \mathbf{j}\{\dot\omega_{(y)} I_{(y)} - \dot\omega_{(z)}\textstyle\int_m yzdm - \dot\omega_{(x)}\int_m xydm\}; \\ \mathbf{M}'_{(z)} &= \mathbf{k}\{\dot\omega_{(z)} I_{(z)} - \dot\omega_{(x)}\textstyle\int_m zxdm - \dot\omega_{(y)}\int_m yzdm\}; \end{aligned}\right\} \quad (89)$$

in which the same six integrals occur that the choice of principal axes eliminates. Consequently if we use at the epoch the projections upon the principal axes, we obtain

$$\mathbf{M}' = \dot\omega_{(x)} I_{(x)} + \dot\omega_{(y)} I_{(y)} + \dot\omega_{(z)} I_{(z)}. \quad \text{[Principal axes.]} \quad (90)$$

This adds one feature to the previous conclusion in equation (84), and makes evident that (**M**′) cannot vanish while ($\dot{\boldsymbol\omega}$) differs from zero, as a limitation upon the subtractive element of equation (82). And it throws stronger light upon a possible constancy of (E) while both (**M**′) and (**M**″) are active, for which the condition is that (**M**′) as well as (**M**″) should be perpendicular to ($\boldsymbol\omega$). This is compatible with the presence of ($\dot{\boldsymbol\omega}$) since the latter may have any direction relatively to ($\boldsymbol\omega$). Where the time-derivatives of equations like (80) play a part in such considerations as the foregoing, of course it may be necessary to take account of variable moment of inertia as being important in reconciling the presence of ($\dot{\boldsymbol{\omega}}$) and the absence of (**M**).

Should the rotation that is under investigation be about the center of mass of the body, the force to be brought in for the accompanying translation or to accelerate the particle of the combination is calculable as ($m\dot{\bar{\mathbf{v}}}$), where any value may have been assigned by other elements to the second factor. But if the case is one of pure rotation round any origin or fixed point, it is plain that the acceleration and velocity of the center of mass are prescribed at the values

$$\dot{\bar{\mathbf{v}}} = (\dot{\boldsymbol{\omega}} \times \bar{\mathbf{r}}) + (\boldsymbol{\omega} \times \bar{\mathbf{v}}); \qquad \bar{\mathbf{v}} = (\boldsymbol{\omega} \times \bar{\mathbf{r}}), \tag{91}$$

requisite locally under the rule of equations (44, 72). Then the total force brought to bear must be accurately adjusted to produce this acceleration, and a constraint at the origin may have to be made active in order to give exactly the requisite force. For reasons of that nature, the constraint may need to be calculated or expressed, although it can contribute nothing to the moment (**M**) about the origin, and can in that respect be ignored. It rests upon the general understanding about sections 45 and 51, that all the leading equations like (86, 88, 89) are adaptable to center of mass as origin without formal change, and by mere substitution of the values then effective.

CHAPTER III

Reference-Frames: Transfer and Invariant Shift

64. Let us recall now the fact that the exercise of choice of reference-frame must be an assumed preliminary to determining any definite working values for the fundamental quantities, and consequently for all quantities calculable in terms of them. This is not interfered with as a truth by our predominant habit of making the earth's surface locally the tacitly adopted basis of reference. The circumstances then bring with them quite naturally a recognizable need of deliberately guided inquiry into the extent to which such values are affected by an allotted range in selection and specification for our reference-frame. This will afford the necessary machinery for correct transfer from one reference-frame to another as standard when that is dictated by an effort at greater precision or by reasons founded in an advantage of convenience.

The line of thought to be taken up next will trace out those matters of material consequence connected with the chief kinematical and dynamical expressions which require for their settlement a collation of values resulting when particular frames are chosen among a group that are in assigned conditions of relative configuration and motion. The fullest survey belonging to that discussion embraces much that would be scarcely relevant on the scale laid down for our present undertaking. But by allowing the more practical interests in these directions to set the limits, we shall confine our scope to methods that are in most frequent use for translating the important expressions into convertible terms of familiar type and ascertaining their mutual dependence. In so far as vectors can be made the vehicle of expression, they are likely to deal directly with resultants and totals, and then we are concerned with the amounts by which these change at a transfer from one frame to another. Yet because we must at times prepare more completely for computation, this alone would constrain us to sacrifice to those ends the compactness of vectorial statements. Other reasons also compel us to find place for the partials or components that are characteristic of various coördinate systems whose peculiar advantages make them useful auxiliaries to the reference-frame; and this will raise a second group of questions. Some close intrinsic connections will be found, however, to make interdependent the two branches of the inquiry, relating one to the uses of coördinate systems and the other to comparisons among reference-frames, which occupy this chapter and the next.

65. First as to transfers and comparisons among reference-frames. Since scalar mass that is unaffected by position and motion becomes by that supposition neutral to the main issues here, something can be done toward clearing the ground by noticing at once how many important decisions must then turn upon the kinematical factors; solely upon these in the differential elements, though as we have found at certain points in the preceding chapter, the mass-distribution continues to play some part through the integrals that are related to the center of mass and to the moments of inertia. Accordingly we are enabled to restrict ourselves in the first steps to kinematics, essentially to radius-vectors and velocities and accelerations, the properly dynamical phase being covered finally by introducing the necessary mass factors.

As one aid to brevity, we shall outline a notation by way of preface, to be used consistently throughout the combinations and comparisons that we must make. Let one reference-frame established by its origin (O) and its axes (XYZ) be constituted the standard, the axes being orthogonal and in the cycle of a right-handed screw. By affording to our thought one term common to a series of comparisons, this frame will furnish a means of coördinating their individual results. Let any one of the other reference-frames with which we may happen to be concerned alternatively, either under suggestion from special conditions or for the purpose of more general discussion, be determined through its origin (O′) and its axes (X′Y′Z′) and be distinguished as a comparison-frame. All the frames are supposed congruent. We shall preserve a helpful symmetry of notation by assigning regularly primed quantities to comparison-frames and unaccented symbols to the standard. But we must not fail to remember either that the distinction which sets off one frame as standard is for convenience of correlation only, in the first instance, and it retains its arbitrary element until physical reasoning can be seen to converge noticeably or convincingly upon one frame, or a set of frames meeting formulated conditions, as the basis better accommodated to the ultimate statement of any physical laws or regular sequences among phenomena. We have touched on this point in sections 6 and 7. In the preliminary view every frame is qualified for selection to be standard, in relation to which all the others fall into their status of comparison-frames.

66. The configuration of any (O′, X′Y′Z′) relative to the standard can be specified as though it had arisen in virtue of a displacement from original coincidence with (O, XYZ), without needing to imply, however, that the coincidence once existed in reality and that the final configuration has developed progressively by a time process, but also without excluding the latter possibility. In order to dispose of certain aspects of the matter, let us at first conceive definitely all these individual configurations to be permanent, each comparison-frame being taken in a configuration that it retains. Then any continuous transitions within an arrangement of such frames will associate themselves rather with grouping it into a space locus, and no idea will be imported into it of those other features belonging distinctively to motion and a path. But we must expect to find here as elsewhere, that the two points of view run easily one into the other, with those groups of virtual displacements, indicated as possible without violating the conditions for the locus, becoming an actual series in time when the paths are described. One moving frame can mark the positions of all members of a group that are in permanent configurations, as it coincides with them in succession. In point of fact, several similar modulations of the thought here hinge alike upon that dual conception of the elements that enter.

67. The assignment of its relative configuration will involve in general for any frame both a difference of position between (O′) and (O) and a difference of orientation between (X′Y′Z′) and (XYZ). Moreover these two data are assignable independently, and it is intuitively true that the actual localization of (O′, X′Y′Z′) is reproducible from coincidence with (O, XYZ) by combining them in either order. Let the *parallel displacement* or translation of the axes with the origin (O′) be specified by the vector (OO′) which we shall denote by ($\mathbf{r}_0$). And the changed orientation is equivalent to a subsequent displacement by rotation of (X′Y′Z′) as a rigid cross, because they are congruent with (XYZ) and remain orthogonal. Using the notation of section 45, we can indicate the result by the vector sum

$$\gamma = \int d\gamma, \tag{92}$$

with the possibility attaching to resultants in general, of representing equivalently many sets of components.

If the idea of succession enters the last equation, the present connection confines it to a timeless series of elements ($d\gamma$), in each of which the constituents ($\lambda\mu\nu$) or substitutes for them are coexistent. Where it will not cause confusion, the term rotation-vector can be applied to ($d\gamma$), as well as to ($\dot{\gamma}$) of the earlier section. For any comparison-frame accordingly its configuration is given with the requisite definiteness by the two total displacements taken in either order,

$$\mathbf{r}_0 = \int d\mathbf{r}_0; \qquad \boldsymbol{\gamma} = \int d\boldsymbol{\gamma}. \tag{93}$$

68. Let us introduce next any point (Q) having at a given epoch radius-vector ($\mathbf{r}$) in the standard, and ($\mathbf{r}'$) in a comparison-frame. The difference of orientation alone while (O′) coincided with (O) would leave the radius-vector invariant for all permissible sets of axes, the expression of which condition can be put into terms of the two sets of unit-vectors,

$$\mathbf{r} = \mathbf{i}x + \mathbf{j}y + \mathbf{k}z = \mathbf{i}'x' + \mathbf{j}'y' + \mathbf{k}'z'; \tag{94}$$

where the invariance is noticeably obscured until the vector algebra brings it into full relief. The alternative relation accompanying separation of (O′) and (O) is

$$\mathbf{r} = \mathbf{r}_0 + \mathbf{r}', \tag{95}$$

whose form obviously excludes equality of ($\mathbf{r}$) and ($\mathbf{r}'$) so long as ($\mathbf{r}_0$) differs from zero. It should be observed about the last equation that it is based rather upon a triangle as graph than upon a parallelogram, because the conception of ($\mathbf{r}'$) makes it a localized vector with (O′) for base-point.

Regarding now (Q) as typical in any continuous or discontinuous assemblage of points, and (Q′) as any other such point whose radius-vectors in the two frames appear in the allowable forms ($\mathbf{r} + \Delta\mathbf{r}$), ($\mathbf{r}' + \Delta\mathbf{r}'$), we have for the vector (QQ′)

$$\Delta\mathbf{r} = \Delta\mathbf{r}', \tag{96}$$

throughout the group of points, independently of the points chosen and of the particular comparison-frame employed. This records the patent truth that the arrangement of members in any point-group, or their configuration relative to each other, is expressible invariantly by means of the standard frame and of every (O′, X′Y′Z′). With that meaning the remark is to be accepted that " Position coördinates appear in our equations by a convenient fiction only, they being parasitic and auxiliary variables that can be eliminated."[1]

69. If for sufficient reason we maintain the discrimination between (Q) and (Q′) as two individual points and locate each permanently in its configuration with (O, XYZ); or let each be *fixed in the space attached to the standard reference-frame* in the words of one current phrase; no questions about time-derivatives of (**r**), (**r**′), (Δ**r**) or (Δ**r**′) can arise, so long as the configuration of (O′, X′Y′Z′) is also by supposition permanent. The source of those reasons and their cogency will depend upon the case in hand; they may be physical in their nature and extracted by interpretation and analysis from observation, or their origin may be frankly due to a feature in the mathematical treatment. By associating other such individual points with (Q) and (Q′) we may build up a continuous group as a limit, for which the general radius-vector becomes in length a function of its orientation but the essentials of the description remain timeless.

However in any unforced survey of other particular circumstances and their plain suggestions a competitive view must find recognition, that will regard both (Q) and (O′, X′Y′Z′) as individuals somehow identifiable through a series of changing configurations in (O, XYZ), and consequently any account that aims at practical completeness cannot neglect coördinating the two alternatives. There is the elementary fact, for example, that the same dependence of radius-vector upon its orienting angle as before can be presented with both variables made functions of time. But the fruits of that idea are not exhausted in one announcement at the threshold of the matter. For when in our view (Q′) becomes a *subsequent position* of the point (Q), or whenever, more inclusively, the varying position of a moving point is matched at each

[1] Quoted from Poincaré.

epoch with the permanent position of a coincident point, the twofold relation of the same symbols to which this leads with such a double point will reappear perpetually. This can make either aspect of the coincidence a continuous indicator or marker for the other, by means of some connecting rule that formulates from either side the relation of consecutive values—here of the radius-vector. Neither phase of the combination can be ignored or subordinated, without losing hold upon ideas that are central in evaluating any variable quantity by legitimate transition to a substituted uniform condition.[1]

70. These considerations confront us with the necessity of preparing here for that kind of transition, and conceiving (Q) and (O′, X′Y′Z′) to be individual and moving. This can be executed conveniently by subdividing into steps, and taking first the one that affects (Q) alone, while we retain for the time being that permanent configuration of (O′, X′Y′Z′) in the standard frame which is afterwards to be abandoned. If we accept for (Q) and (Q′) a fusion of identity in the sense that they are now adopted as two positions of the same moving point, terminal for any time-interval (Δt), the mean velocities for that interval will be equal in our two reference-frames, and also the instantaneous velocities at the epoch beginning the interval. This conclusion finds expression in sequence with the requisite new reading of equation (96) as

$$\mathbf{v} \equiv \mathrm{Lim}_{\Delta t=0}\left(\frac{\Delta \mathbf{r}}{\Delta t}\right) = \mathrm{Lim}_{\Delta t=0}\left(\frac{\Delta \mathbf{r}'}{\Delta t}\right) \equiv \mathbf{v}'; \qquad (97)$$

or in semi-cartesian dress,

$$\mathbf{v} = \mathbf{i}\frac{dx}{dt} + \mathbf{j}\frac{dy}{dt} + \mathbf{k}\frac{dz}{dt} = \mathbf{i}'\frac{dx'}{dt} + \mathbf{j}'\frac{dy'}{dt} + \mathbf{k}'\frac{dz'}{dt} = \mathbf{v}', \qquad (98)$$

since both sets of unit-vectors are by supposition constant here, as well as ($\mathbf{r}_0$). And further, because these simultaneous veloci-

[1] See Note 19.

ties of (Q) are thus continually equal vectors, it is an evident corollary that the accelerations of (Q) in the two frames are always equal at the same epoch; or

$$\dot{\mathbf{v}} \equiv \mathrm{Lim}_{\Delta t=0}\left(\frac{\Delta \mathbf{v}}{\Delta t}\right) = \mathrm{Lim}_{\Delta t=0}\left(\frac{\Delta \mathbf{v}'}{\Delta t}\right) \equiv \dot{\mathbf{v}}', \qquad (99)$$

whose expanded equivalent again is

$$\dot{\mathbf{v}} = \mathbf{i}\frac{d^2x}{dt^2} + \mathbf{j}\frac{d^2y}{dt^2} + \mathbf{k}\frac{d^2z}{dt^2}$$
$$= \mathbf{i}'\frac{d^2x'}{dt^2} + \mathbf{j}'\frac{d^2y'}{dt^2} + \mathbf{k}'\frac{d^2z'}{dt^2} = \dot{\mathbf{v}}'. \qquad (100)$$

71. Taken together, these statements make clear for every epoch the invariance of velocity and acceleration that holds good throughout any group of reference-frames that are in permanent relative configuration. Also the consequences in application to the same system of bodies at the same epoch are apparent. Each local velocity and acceleration being unaffected, the six fundamental quantities show in the standard and in any comparison-frame of the group thus correlated:

$$\left.\begin{aligned} &\mathbf{Q} = \mathbf{Q}'; \quad \mathbf{R} = \mathbf{R}'; \quad \mathrm{E} = \mathrm{E}'; \quad \mathrm{P} = \mathrm{P}'; \\ &\mathbf{H} = \mathbf{H}' + (\mathbf{r}_0 \times \mathbf{Q}'); \quad \mathbf{M} = \mathbf{M}' + (\mathbf{r}_0 \times \mathbf{R}'), \end{aligned}\right\} \qquad (101)$$

which it may be well to compare for likeness and difference, say when $(\mathbf{r}_0 = \bar{\mathbf{r}})$, with the corresponding relations exhibited in section 51, the contrast between (C′) there and (O′) here lying in the freedom of the former point to move with velocity and acceleration. The less narrowly limited connection of center of mass with force-moment and moment of momentum should be realized.

72. The foregoing results are sufficiently practical in their bearing to incite us to appropriate, without delaying, the possibilities that they illustrate. These lie in the direction of a certain liberty to employ what amounts to a whole series of different reference-frames at successive epochs of the same problem, or inside the range covered by one discussion, and yet avoid prohibitive complications that might be due to such repeated transfers to new standards. Provided only that we observe those restrictions which underlie the invariance of any particular quantities with which we are dealing, the frames become interchangeable in respect to them; and freedom prevails to depart, at later epochs and as often as may prove desirable, from the initial choice of reference-frame. At least it is evident how there will be no danger, on relinquishing one frame and adopting another subject to the proper conditions, of dislocating ruinously by breaking into it the expression of a continuous series of values for any quantity that the change leaves invariant. Dislocations of minor scope can be reckoned with otherwise, or often disregarded, where they enter.

Such procedure remains clearly valid, always within its limitations, whether its revisions of choice involve configurations separated by steps that are finite or that are made with finite pauses between them, or whether the group of frames used melts at the limit into a continuously consecutive arrangement. It is equally permissible, besides, to regulate the employment of members in a group of frames according to a time-schedule, or to effect timeless transitions among frames and to concern ourselves comparatively with simultaneous values of different quantities, or finally of the same quantity when we break the barrier of invariance. The actual working out of the main thought rings the changes on all these offered chances, so that several of the combinations will come before us prominently for specific examination.

73. We proceed next to remove the limitation that has held us to permanent configuration for (O′, X′Y′Z′). We relax this permanence relative to (O, XYZ) by admitting, first changes in ($\mathbf{r}_0$) alone while (γ) is unchanging in equation (93), and afterwards the full freedom with changes in (γ) also. It seems advantageous to attack this phase of the matter, too, through what we have spoken of as fusion of identity; but now for comparison-frames that like the points (Q) and (Q′) can from another approach also be distinguished as separate individuals. Return then to that original view of those points, include some second comparison-frame (O″, X″Y″Z″) and carry on the notation by adding

$$O'O'' \equiv \Delta\mathbf{r}_0; \qquad O''Q' \equiv \mathbf{r}''. \tag{102}$$

The relations associating (Q) with (O′) and (O), and (Q′) with (O″) and (O) are

$$\mathbf{r} = \mathbf{r}_0 + \mathbf{r}'; \qquad \mathbf{r} + \Delta\mathbf{r} = (\mathbf{r}_0 + \Delta\mathbf{r}_0) + \mathbf{r}'', \tag{103}$$

showing by their difference

$$\Delta\mathbf{r} = \Delta\mathbf{r}_0 + (\mathbf{r}'' - \mathbf{r}'), \tag{104}$$

whose verbal equivalent can be read from the broken line (QO′O″Q′) that is equal as a vector sum to (QQ′) and closes a quadrilateral that may be of course either gauche or plane.

We may now retrace the previous track further, whenever we can attribute to the frames (O′, X′Y′Z′) and (O″, X″Y″Z″) some adequate basis of continuous identity similar to that which was made to unite (Q) and (Q′), so that the entire group of discrete frames of permanent but differing configurations is replaced by the conception of one representative frame (O′, X′Y′Z′) in continuously variable relation to the standard. First, confine attention to the origin (O′), deferring a little the introduction of changing orientation, suppose ($\mathbf{r}_0$) to vary with time and read equation (104) to correspond. The originally unrelated vectors ($\mathbf{r}'$) and ($\mathbf{r}''$) coalesce under one symbol ($\mathbf{r}'$) when that is used to signify a vector drawn always from the position of (O′) at any epoch to the simultaneous position of (Q). It is therefore a vector to be rated in the standard frame as localized, but variable in all three particulars of length, orientation and base-point. In pursuance of that thought write

$$\mathbf{r}'' - \mathbf{r}' = \Delta\mathbf{r}', \tag{105}$$

divide equation (104) by the elapsed time (Δt) and proceed to record the limiting ratio in the form

$$\text{Lim}_{\Delta t=0}\left(\frac{\Delta\mathbf{r}'}{\Delta t}\right) = \dot{\mathbf{r}} - \dot{\mathbf{r}}_0 \equiv \mathbf{v} - \mathbf{v}_0, \tag{106}$$

if ($\mathbf{v}$) and ($\mathbf{v}_0$) denote the velocities of (Q) and of (O′) in the standard frame. The formal repetition in this first member of ($\mathbf{v}'$) as specified in the terms of equation (97) is significant of its unconstrained meaning here too as the velocity of (Q) reckoned in the frame (O′, X′Y′Z′), but under an extension that allows a supposed motion of (O′). Duly observing the imposed condition of unchanging orientation for ($\mathbf{i}'\mathbf{j}'\mathbf{k}'$) that is still maintained, confirm this feature of the development by writing the time-derivative of the permanent relation in equation (95) in the form

$$\mathbf{v}' \equiv \mathbf{i}'\frac{dx'}{dt} + \mathbf{j}'\frac{dy'}{dt} + \mathbf{k}'\frac{dz'}{dt} = \dot{\mathbf{r}} - \dot{\mathbf{r}}_0, \tag{107}$$

and compare with equation (98). It is plain that ($\mathbf{v}'$) and ($\mathbf{v}$) are equal at any epoch when ($\dot{\mathbf{r}}_0$) is zero.

74. These thoughts harmonize in another respect with equation (106) if we see registered there a consequence of a double process of incrementation for the vector ($\mathbf{r}'$), now completely variable in the standard frame, with rate ($\mathbf{v}$) at its forward end and with rate ($\mathbf{v}_0$) at its base-point. In every such combination, so long as these rates are equal, the vector retains its length and orientation in the reference-frame; as a free vector it remains equal at all epochs, though as a localized vector it experiences change of position determined by the common value of the two rates. In the less particularly chosen suppositions where the two rates are unequal, only their difference such as ($\mathbf{v} - \mathbf{v}_0$) is available to give change of tensor and of orientation.

But to take account of these latter elements for ($\mathbf{r}'$) and to ignore or drop out the change in position for (O′) substitutes effectively (O′, X′Y′Z′) as reference-frame, the orientation of ($\mathbf{i}'\mathbf{j}'\mathbf{k}'$) having first and last the requisite permanence, so that the transfer is uncomplicated in that respect. And since the part ($\mathbf{v}_0$) applies simultaneously or in common to all points (Q), the readjustment of velocity values made necessary by this type of transfer to a new reference-frame (O′, X′Y′Z′) can be summarized as the subtraction of a *translation with the velocity of the new origin* in the first standard frame. In connection with this the thought frequently finds expression that each frame carries its space in rigid attachment to it, and these interpenetrating spaces will have in the present case at each coincident pair of points the relative velocity ($\pm\, \mathbf{v}_0$) at any epoch.

The effects upon acceleration of a similar transfer while ($\mathbf{i}'\mathbf{j}'\mathbf{k}'$) remain constant show plainly on forming the time-derivative of equation (107). This gives

$$\dot{\mathbf{v}}' \equiv \mathbf{i}'\frac{d^2x'}{dt^2} + \mathbf{j}'\frac{d^2y'}{dt^2} + \mathbf{k}'\frac{d^2z'}{dt^2} = \dot{\mathbf{v}} - \dot{\mathbf{v}}_0; \quad \dot{\mathbf{v}} - \dot{\mathbf{v}}' = \dot{\mathbf{v}}_0; \quad (108)$$

and the proper allowance shows again in terms of a translation with the new origin (O′), whose acceleration, however, is now essential and not its velocity. In the light of equations (107, 108) the combinations become self-evident by which velocities or accelerations or both may be left invariant under a change of reference-frame. The bearing upon the segregation in sections 21, 31, 48 and 49 will not escape attention.

75. In order to embrace finally the transition to axes (X′Y′Z′) whose orientation is changing in the standard frame, while they are accompanying their origin (O′), we can use our knowledge that the rotation-vector of sections 45 and 67 specifies such changes adequately, and thus complete under the wider play of these conditions the time-derivative of the relation that remains valid,

$$\mathbf{r} = \mathbf{r}_0 + \mathbf{r}' = \mathbf{r}_0 + (\mathbf{i}'x' + \mathbf{j}'y' + \mathbf{k}'z'). \qquad (109)$$

Upon the supposition that the group ($\mathbf{i}'\mathbf{j}'\mathbf{k}'$) are at the epoch varying in direction relative to (XYZ) as determined by the rotation-vector ($\dot{\gamma}$), we are led by the differentiation directly to the equation

$$\dot{\mathbf{r}} = \dot{\mathbf{r}}_0 + (\dot{\gamma} \times \mathbf{r}') + \left(\mathbf{i}'\frac{dx'}{dt} + \mathbf{j}'\frac{dy'}{dt} + \mathbf{k}'\frac{dz'}{dt}\right), \qquad (110)$$

from which it follows that

$$\mathbf{v} - \mathbf{v}' = \mathbf{v}_0 + (\dot{\gamma} \times \mathbf{r}'); \qquad \mathbf{v} = \mathbf{v}' + [\mathbf{v}_0 + (\dot{\gamma} \times \mathbf{r}')]. \qquad (111)$$

Typical special cases under this equation can be decided by inspection. Note the form now taken by the idea of interpenetrating spaces in section 74, connecting it with the general motion of a rigid solid in section 48. The last group of terms in equation (110) must still be recognized as the velocity ($\mathbf{v}'$) of (Q) in (O′, X′Y′Z′), because the transfer to the latter as the standard cancels perforce from admission into ($\mathbf{v}'$) every change in orientation attributable otherwise to ($\mathbf{i}'\mathbf{j}'\mathbf{k}'$), in addition to ignoring changes in the position of (O′).

76. Various equivalent verbal formulations beside those already suggested can be devised for equations (107, 108, 111), that all amount in principle to a superposition of relative velocities or accelerations. And it will be seen how the same idea can be applied repeatedly and can carry us through a chain of transfers to a final result that accumulates in itself all the contributions at its several steps. Remembering that forces are bound to superposition also, as they enter successively with the acceptance of their accelerations into physical status, trace there a line of advance in precision that would parallel our discarding one reference-frame in favor of another.[1] The same possibility of superposition lies open as we go forward from equation (111) to

[1] See Note 20.

consider the similar transfer for accelerations, though the complications soon cut down any advantage of a verbal expression for it.

Formal routine yields for the time-derivative of the general relation in equation (110) or (111) the result

$$\dot{\mathbf{v}} = \dot{\mathbf{v}}_0 + (\ddot{\boldsymbol{\gamma}} \times \mathbf{r}') + 2(\dot{\boldsymbol{\gamma}} \times \mathbf{v}') + (\dot{\boldsymbol{\gamma}} \times (\dot{\boldsymbol{\gamma}} \times \mathbf{r}')) + \dot{\mathbf{v}}', \qquad (112)$$

in which $(\mathbf{r}')$, $(\mathbf{v}')$, $(\dot{\mathbf{v}}')$ specify the position, velocity and acceleration of any point (Q) by means of (O′, X′Y′Z′); that is, to recapitulate,

$$\mathbf{r}' \equiv \mathbf{i}'x' + \mathbf{j}'y' + \mathbf{k}'z'; \qquad \mathbf{v}' \equiv \mathbf{i}'\frac{dx'}{dt} + \mathbf{j}'\frac{dy'}{dt} + \mathbf{k}'\frac{dz'}{dt};$$
$$\dot{\mathbf{v}}' \equiv \mathbf{i}'\frac{d^2x'}{dt^2} + \mathbf{j}'\frac{d^2y'}{dt^2} + \mathbf{k}'\frac{d^2z'}{dt^2}; \qquad (113)$$

$(\ddot{\boldsymbol{\gamma}})$ is the angular acceleration belonging at the epoch to the rotation-vector $(\dot{\boldsymbol{\gamma}})$, and $(\dot{\mathbf{v}}_0)$ denotes the acceleration of (O′) in (O, XYZ). Interest will center here upon the terms affected by the rotation, into which the elements $(\mathbf{r}')$ and $(\mathbf{v}')$ individual to the point (Q) enter; and for the latter, the connections shown in equation (111) must be duly heeded. It will cultivate control of details in the method to carry through its application to such combinations as $(\dot{\boldsymbol{\gamma}} = 0)$, $(\ddot{\boldsymbol{\gamma}} = 0)$, separately or conjointly, in preparation for the summary that follows. And then to work out lists, comparable with that in section 71, for the general conditions of equations (107, 111, 112), showing how the different quantities are affected by the transfers from one reference-frame to another that have been brought under review. It is always a reciprocal interdependence that is in question, and a procedure for transfer in either direction.

77. To round out this stage of the inquiry, we can now formulate for velocity and acceleration the suppositions necessary to their invariance, that will put the frames for which these are satisfied to that extent on an equal or indifferent footing. We begin with acceleration, whose invariance necessitates conformably to equation (112),

$$\dot{\mathbf{v}}_0 + (\ddot{\gamma} \times \mathbf{r}') + 2(\dot{\gamma} \times \mathbf{v}') + (\dot{\gamma} \times (\dot{\gamma} \times \mathbf{r}')) = 0. \quad (114)$$

But ($\mathbf{v}_0$), ($\ddot{\gamma}$) and ($\dot{\gamma}$) are to be assumed independently of each other; and further, the search is for a general relation covering all points (Q) in all phases of their motion, which puts aside as insufficient every particular adjustment or singular value like

$$\mathbf{r}' = 0; \qquad \mathbf{v}' = 0;$$

or colinear factors in some individual vector products. Hence the proposed invariance of acceleration demands all three conditions,

$$\dot{\mathbf{v}}_0 = 0; \qquad \dot{\gamma} = 0; \qquad \ddot{\gamma} = 0. \quad (115)$$

These permit the comparison-frame to have unaccelerated translation with (O′), but forbid changes in orientation (γ) as indicated by its time-derivatives of the first and second order.

The invariance of velocity imposes different limitations derivable by inspection from equation (111) as being

$$\mathbf{v}_0 = 0; \qquad \dot{\gamma} = 0. \quad (116)$$

The second of these conditions, therefore, is common to the invariance of velocity and of acceleration. But as regards the translation with (O′) equation (116) excludes any velocity ($\mathbf{v}_0$) though allowing an acceleration ($\dot{\mathbf{v}}_0$), while equation (115) inverts these relations. The double condition for invariance of velocity bars at the epoch motion of (O′, X′Y′Z′) in (O, XYZ), but gives freedom as to subsequent states. The triple condition for invariance of acceleration maintains the exclusion of changing orientation and sharpens it by ($\ddot{\gamma} = 0$), but allows any constant value of the vector ($\mathbf{v}_0$).

The above conclusions coupled with the discussion that led

to equations (98) and (100) bring out how (O′, X′Y′Z′) if treated as moving in the standard frame must always sacrifice in some degree the invariant properties in regard to velocity, acceleration and the dynamical quantities dependent upon them; though these are, nevertheless, preserved intact by a succession of frames, each in coincidence with the moving frame at one epoch. The permanent values of ($\mathbf{r}_0$) and (γ) for the stationary frames are marked off, one by one, in the series of instantaneous values for those elements belonging to the moving frame. In this sense and to this extent, the presence or absence of an invariance that happens to be in question can be made to turn upon the point of view, which because it affects values also raises issues that need to be decided in the light of clear statement of the position our thought has occupied. Consequently it is likely to repay us, if we enforce this main idea by approaching it in reliance upon the frames of permanent configuration, the mathematics being modified to match.

Invariant Shift.

78. Whereas the radius-vectors ($\mathbf{r}$) have been handled in the preceding equations as functions of time alone, directly in (O, XYZ) and in (O′, X′Y′Z′) through the relation

$$\mathbf{r} = \mathbf{r}_0 + \mathbf{r}', \tag{117}$$

this second mode of making a beginning will disguise the same radius-vectors ($\mathbf{r}$) into functions of three independent variables (t, $\mathbf{r}_0$, γ). And this will evidently lead toward fixing attention upon a whole group of comparison-frames inclusively, to be constructed by assigning continuous, but otherwise arbitrary, values to ($\mathbf{r}_0$) and (γ), perhaps in connection with equation (93), while (t) remaining unchanged gives simultaneous currency to those values.

The exact differential of ($\mathbf{r}$) indicated according to the new terms is

$$dr = \frac{\partial r}{\partial t} dt + \frac{\partial r}{\partial r_0} dr_0 + \frac{\partial r}{\partial \gamma} d\gamma. \tag{118}$$

This form might indeed be denominated rather sterile of meaning in respect to ($\mathbf{r}$) itself, for it is apparent enough from many of the expressions that we have been laying down that ($\mathbf{r}$) is not intrinsically dependent on either ($\mathbf{r}_0$) or (γ). Similarly if we use equation (117), and after omitting the terms that are necessarily zero, on our assumption about independent variables, write

$$dr = \left(\frac{\partial r_0}{\partial r_0} + \frac{\partial r'}{\partial r_0}\right) dr_0 + \frac{\partial r'}{\partial \gamma} d\gamma + \frac{\partial r'}{\partial t} dt, \tag{119}$$

appeal to equation (94) seems to tell that ($\mathbf{r}'$) at any epoch does not change with (γ). But after admitting that

$$\frac{\partial r}{\partial r_0} = \frac{\partial r_0}{\partial r_0} + \frac{\partial r'}{\partial r_0} = 0; \qquad \frac{\partial r'}{\partial \gamma} = 0; \tag{120}$$

equation (119) is found, notwithstanding, really helpful for the end sought, as a starting-point for collating different sets of components within our group of frames, though it might be superfluous did we restrict ourselves to resultants. In order to develop this idea more fully introduce the semi-cartesian equivalent

$$\mathbf{r}' = \mathbf{i}'x' + \mathbf{j}'y' + \mathbf{k}'z', \tag{121}$$

whose second member is intended for a comprehensive notation applying both tensors ($x'y'z'$) and unit-vectors ($\mathbf{i}'\mathbf{j}'\mathbf{k}'$) generically to the whole group. They are then variables as affected by passage from one frame to its neighbors, and in addition the tensors are variable with time in the same frame.

This temporary identity of the variables in the one frame, which may pick that one out or enable us to recognize it, and yet be evanescent for the group of frames as a whole, lies close to the heart of the thought in equation (119), as contrasted with a completer convection of identity with one moving frame, whose tensors and unit-vectors are consequently functions of time only. For the present purpose, on the other hand, and in its adapted mathematics, the tensors (x′y′z′) must be considered functions of ($\mathbf{r}_0$), (γ), (t); but the unit-vectors ($\mathbf{i}'\mathbf{j}'\mathbf{k}'$) and ($\mathbf{r}_0$) do not at this stage vary by mere lapse of time; nor the former by relocation of the origin (O′)—they must be functions of (γ) alone. Under the suppositions and the reasons for them thus made explicit, we execute the differentiation of equation (121) in combination with equation (119) and obtain

$$\begin{aligned} d\mathbf{r} = \frac{\partial \mathbf{r}_0}{\partial r_0} dr_0 &+ \left(x' \frac{\partial \mathbf{i}'}{\partial \gamma} + y' \frac{\partial \mathbf{j}'}{\partial \gamma} + z' \frac{\partial \mathbf{k}'}{\partial \gamma} \right) d\gamma \\ &+ \mathbf{i}' \left(\frac{\partial x'}{\partial r_0} dr_0 + \frac{\partial x'}{\partial \gamma} d\gamma + \frac{\partial x'}{\partial t} dt \right) \\ &+ \mathbf{j}' \left(\frac{\partial y'}{\partial r_0} dr_0 + \frac{\partial y'}{\partial \gamma} d\gamma + \frac{\partial y'}{\partial t} dt \right) \\ &+ \mathbf{k}' \left(\frac{\partial z'}{\partial r_0} dr_0 + \frac{\partial z'}{\partial \gamma} d\gamma + \frac{\partial z'}{\partial t} dt \right). \end{aligned} \tag{122}$$

79. This expansion supplies material to interpret profitably, when it is observed that the imposed condition for the partial time-derivatives with the set of variables now adopted is the same in effect as that for invariant velocity to which equation (97) is subject. Consequently the three terms on the left are properly equated to the velocity of any (Q) in the standard frame, when we write

$$\mathbf{i}' \frac{\partial x'}{\partial t} + \mathbf{j}' \frac{\partial y'}{\partial t} + \mathbf{k}' \frac{\partial z'}{\partial t} = \mathbf{v}. \tag{123}$$

The double use of this equality is apparent, either in obtaining projections of known ($\mathbf{v}$) upon the (X′Y′Z′) of the configuration, or in determining ($\mathbf{v}$) by means of its projections upon whatever particular comparison-frame is designated by the stationary values at which ($\mathbf{r}_0$) and (γ) are arrested while the partial change with (t) is recorded.

Thus no essential in regard to consistent expression of velocities would be sacrificed if we depended upon any such comparison-frame momentarily to replace (O, XYZ) in its service as standard, and did likewise for new stationary values of ($\mathbf{r}_0$) and (γ) with velocities at other epochs. This comment will infuse its due quota of meaning into the equality

$$\dot{\mathbf{r}} \equiv \mathbf{v} = \frac{\partial \mathbf{r}'}{\partial t} \tag{124}$$

and parallel expressions, whenever similar opposed total derivatives and partials are made to play their rôles as the basis of a regular procedure, in which a resultant vector is to be constructed or evaluated by means of components parallel to axes that differ systematically, or in which the projections of a given vector upon such axes appear naturally.

It is readily apprehended, at this point, how such plans are effectively equivalent to a continuous process of transfer to new standard frames that is kept simple by its preservation of invariance, while it may secure a permanence of form or other advantage in addition. The indispensable resolution of acceleration along tangent and normal of the epoch in treating curved paths is one case in point; and the compact forms obtained by introducing principal axes will suggest strongly some similar scheme in continuation of sections 61 and 63 with expectation of profit from it. It seems convenient to have a brief name for contrived plans of this character, so we shall refer to them hereafter as *shift* of reference-frame, implying always invariant shift in so far as some quantities are not thereby modified from the simultaneous value indicated in the standard frame.[1]

80. The three terms put down in equation (123) are then seen to reproduce accurately in the combinations of equation (122) the actual displacement (d**r**) for the time (dt) of the moving point (Q) in the standard frame; and therefore, the

[1] See Note 21.

remaining entries in the coefficients of ($\mathbf{i}'\mathbf{j}'\mathbf{k}'$) must be illusory if taken by themselves, as regards describing what is thus happening at (Q). In fact, as their form involving constancy of (t) indicates clearly, they are attendant upon comparisons of corresponding and simultaneous pairs in two sets of projections determining or determined by the same ($\mathbf{r}'$), but connected with two sets of axes differing in orientation by ($d\boldsymbol{\gamma}$) and having origins separated by ($d\mathbf{r}_0$). The complete coefficients of ($\mathbf{i}'\mathbf{j}'\mathbf{k}'$) being evidently the exact differentials for the present independent variables of the tensors ($x'y'z'$), equation (122) can be rewritten

$$d\mathbf{r} = d\mathbf{r}_0 + (d\boldsymbol{\gamma} \times \mathbf{r}') + (\mathbf{i}'dx' + \mathbf{j}'dy' + \mathbf{k}'dz'), \qquad (125)$$

if we bring in the consequences of the rotation-vector ($d\boldsymbol{\gamma}$) in the form

$$\left(\frac{\partial \mathbf{i}'}{\partial \gamma} + \frac{\partial \mathbf{j}'}{\partial \gamma} + \frac{\partial \mathbf{k}'}{\partial \gamma}\right) d\gamma = d\boldsymbol{\gamma} \times (\mathbf{i}' + \mathbf{j}' + \mathbf{k}');$$

$$\left(x'\frac{\partial \mathbf{i}'}{\partial \gamma} + y'\frac{\partial \mathbf{j}'}{\partial \gamma} + z'\frac{\partial \mathbf{k}'}{\partial \gamma}\right) d\gamma = d\boldsymbol{\gamma} \times \mathbf{r}'. \qquad (126)$$

Accordingly equation (125) in its second member is so arranged that it includes within its last group deviations from the true value of ($d\mathbf{r}$) through apparent or spurious changes in the tensors, and finally offsets these by the corrective first and second terms.

That exactly the compensating adjustment shown must exist, can be argued summarily, in line with our remark upon equations (119, 120), from the independence of actual changes in ($\mathbf{r}$) of mere subheadings in our accounts of them, but some few details are worth inserting for emphasis. The first of equations (120) is self-evident, for ($\mathbf{r}'$) must lose whatever ($\mathbf{r}_0$) gains, while ($\mathbf{r}$) is held at its value by unchanging (t). Let us therefore analyze only the second of those equations in regard to the dependence of the tensors upon (γ). We must have

$$x' = \mathbf{i}'\cdot(\mathbf{x} - \mathbf{x}_0) + \mathbf{i}'\cdot(\mathbf{y} - \mathbf{y}_0) + \mathbf{i}'\cdot(\mathbf{z} - \mathbf{z}_0). \qquad (127)$$

Then because neither (xyz) nor ($x_0y_0z_0$) in the standard frame are dependent upon (γ),

$$\begin{aligned}\frac{\partial x'}{\partial\gamma}d\gamma &= \left(\frac{\partial \mathrm{i}'}{\partial\gamma}d\gamma\right)\cdot((\mathbf{x} - \mathbf{x}_0) + (\mathbf{y} - \mathbf{y}_0) + (\mathbf{z} - \mathbf{z}_0)) \\ &= \left(\frac{\partial \mathrm{i}'}{\partial\gamma}d\gamma\right)\cdot\mathbf{r}'.\end{aligned} \qquad (128)$$

Consequently

$$\frac{\partial x'}{\partial\gamma}d\gamma = (d\boldsymbol{\gamma}\times\mathbf{i}')\cdot\mathbf{r}' = -(d\boldsymbol{\gamma}\times\mathbf{r}')\cdot\mathbf{i}'; \qquad (129)$$

and similarly

$$\frac{\partial y'}{\partial\gamma}d\gamma = -(d\boldsymbol{\gamma}\times\mathbf{r}')\cdot\mathbf{j}'; \qquad \frac{\partial z'}{\partial\gamma}d\gamma = -(d\boldsymbol{\gamma}\times\mathbf{r}')\cdot\mathbf{k}'; \qquad (130)$$

which together prove consistently with anticipation,

$$\left(\mathrm{i}'\frac{\partial x'}{\partial\gamma} + \mathrm{j}'\frac{\partial y'}{\partial\gamma} + \mathbf{k}'\frac{\partial z'}{\partial\gamma}\right)d\gamma = -(d\boldsymbol{\gamma}\times\mathbf{r}'). \qquad (131)$$

81. Let us next return to equation (110), with the reminder that it occurs in a general procedure of substituting a new reference-frame to be standard, by making necessary allowance for the relative motion of the two frames. Multiply both members by (dt) and verify that its form then becomes identical with equation (125), although the latter was deduced under more special limitations that we propose to distinguish as shift, and that keep the velocities invariant. In other words, the sum of the last three terms in this equation will differ by the same amount from an actual displacement (dr) in the standard frame, whether ($d\mathbf{r}_0$) and ($d\boldsymbol{\gamma}$) designate differentially changes of configuration observable in the one moving comparison-frame, or whether the same elements express the shift in passage to a consecutive member of the invariant group of frames.

These two relations distinct in their conceived source are joined into a formal identity, primarily because together they embrace a series of coincidences, as displayed in sections 69 and 77, for each aspect of which the same symbols can be given coherent meaning. But that fact though patent is no good ground for obliterating either one of the serviceable conceptions out of which the equation that we are now discussing has arisen for us. We should rather grasp firmly the thought that two successions are here instructively coördinated: one ensuing by movement of an identified frame into new positions, and the other by timeless shift to new stationary frames. These conclusions refer in this first instance, of course, only to the velocities for which they have been established; but they are conveniently capable of extensions. In the measure that these are unfolded, they will lend finally to the otherwise trivial identity

$$\mathbf{A} = (\mathbf{A} - \mathbf{B}) + \mathbf{B} \tag{132}$$

that equation (125) may suggest, a value for working needs through practically advantageous selections of (**B**). Note, for example, that equation (74) is scarcely different in type.

82. As the last remark might imply somewhat plainly, the exploitation of the dominating idea in shift will look to govern its course and its extent by special phases of adaptation contrived to meet combinations that do occur. Analysis that we shall undertake of several coördinate systems may be expected to illustrate and repeat that lesson. What the instances quoted in section 79 show is more generally true: That the plans for shift require various adjustments to be renewed continuously, and keep modulated pace with conditions that develop velocity, acceleration and the closely related dynamical quantities. Thus the progress of the shift must accommodate itself to a regulative time-series of other values, and this in turn imposes upon the shift process itself a necessary rate in time. That situation the mathematics handles by recognizing ($\mathbf{r}_0$) and (γ) to be functions of time, instead of treating them as independent variables subject only to timeless change; so linking them with each other and with the salient phenomena that are to be followed up that some line of advantage sought is most nearly secured.

Nevertheless since the previously independent increments still form a background, these additional functions of time will differ in certain respects from those that yield, for instance, the velocities and accelerations of the moving points (Q). One formulation of the critical difference declares that the latter class of time functions is dictated altogether by an objective element; they must conform to the phenomena studied and express them, their own nature and form being to that important extent not under control. Those of the former class are open to free choice, although we may grant, indeed, that this control is exercised normally in bringing to pass some mode of subordination to what is occurring in other sequences, to the end of attaining simpler models in equations, or the like removal of complications. This employment of time functions in dynamics that are distinguishable in their nature, has long been commented upon and provided for, though the discrimination is stated variously and not always in clearest terms.[1]

On a foundation of the foregoing explanation or some equivalent, we are brought to accept a two-fold dependence upon time in equation (122) and in any statements that disclose to examination the grounds for a similar distinction. Thus we gain the liberty to regard the partial processes as simultaneous, to divide equation (125) by (dt) and so to establish an exact formal identity with equation (110) by allowing for shift rates that are independently assignable. Yet the alternative readings diverge still in the direct meanings associated with ($\dot{\mathbf{r}}_0$) and ($\dot{\gamma}$); these are alike, however, in standing equally among the controllable time

[1] See Note 22.

rates, because the one definite frame to which transfer shall be executed may move at will, save as outlook toward convenience guides or special circumstances demand. Perhaps it is not over-subtle either to insist upon a second residual difference: The plan of equation (110) aims primarily to connect properly two sets of values for velocity, each correct and complete for its own conditions; but equation (125), on the contrary, entertains only one set of values as correct, that are made to reappear finally from being obscured under a transient distortion.

83. We should not have elaborated these ideas with equal fullness had the results borne solely upon the narrower issues gathered about the radius-vector, and had not Hamilton's hodograph given a clew toward making the radius-vector representative of other vectors, and the velocity of its extremity a key to the general vector's time rate. The vector algebra having fallen heir to these methods and enlarged them, it is natural to look upon the previous section as a preface and proceed to trace again its characteristic connections when any vector ($\mathbf{V}$) has replaced ($\mathbf{r}$), and its time-derivatives are offered in parallel with ($\mathbf{v}$) and ($\dot{\mathbf{v}}$). In the course of such extension, we may expect correspondences and fruitful grafting of larger ideas upon the parent special case, all along the line of development whose details are now fairly before us.

But when we come to examine and sort the material that presents itself under such headings, we find the two chief operations that we have been comparing very unequally represented in practice. The circumstances of unrestricted change from one reference-frame to another do reappear in connection with all physical vectors and other types of quantity; and as we have seen exemplified repeatedly already, those changes when they are made necessitate a deliberate reconsideration of all these quantitative values. Yet besides, the occasions that compel such revisions are, at once, comparatively rare and apt to be made for conditions that have become more strongly specialized; although the process is important as regards flawless execution, it shows few features that give it the weight of a procedure that holds its place among the routine methods of frequent use.

The alternative conception that we call shift, however, has been introduced and given preliminary analysis here to a degree that may seem not quite called for, because in the first place it is implicitly or explicitly involved when a number of the standard coördinate systems in dynamics are employed, which is a routine procedure; and because secondly, there has been some failure in clear apprehension and announcement of just those consequences of the restrictions upon the process of shift that bring it into close alliance with the prevailing purpose of coördinate systems. For these are, in the main, adapted to the one central idea of expressing equivalently or invariantly, through some convenient dissection into parts, a resultant or total quantity that relations in a standard frame have first actually or potentially settled upon. When therefore we dismiss in a few sentences the subject of changing reference-frame for the general vector ($\mathbf{V}$), and yet expand the idea of shift on its broader lines, the explanation is to be sought in the reasons that have just been given.

84. If we look again at equation (95) with a view to generalizing upon it, we must describe ($\mathbf{r}_0$) as the difference between the values in the two frames of the vector that is under consideration. Similarly if we write the equation

$$\mathbf{V} = \mathbf{V}_0 + \mathbf{V}' \tag{133}$$

in beginning an attempt to extend the validity of previous conclusions, it is clear how ($\mathbf{V}_0$) is to be read. It is also apparent, or verified by easiest trial, that one obstacle to indicating here a more general rule for change of reference-frame enters because the value of ($\mathbf{V}_0$) depends upon the quantity represented by ($\mathbf{V}$), as instanced by the conditions for invariance in section 77. But it was also forced upon our attention, from equation (94) onward, that ($\mathbf{r}'$) in the standard frame is invariantly given by all frames whose origin is at (O$'$) in its position for the epoch. And while this too draws the lines closer for ($\mathbf{V}'$) and limits narrowly the usefulness of results attached to derivatives of ($\mathbf{r}$), ($\mathbf{r}_0$) and ($\mathbf{r}'$), in doing that it points convincingly toward the process of shift, if we are to generalize, in which this very invariance has been made a prominent characteristic. When we look at the matter from another side, and observe how near an assigned behavior of ($\mathbf{i}'\mathbf{j}'\mathbf{k}'$) comes to furnishing completely the compensating or corrective elements in an equation like (125), once more the conformity of a coördinate system to some rule of displacement can be seen. Thus polar coördinates are essentially a shifting orthogonal set, and a scrutiny of the standard expressions for the components there shows that they meet ($\mathbf{r}'$) on an equal footing of reproducing a resultant invariantly.

85. We shall begin the definite inquiry about shift in its larger relation to coördinate systems by supposing that we have to do with any free vector determined in the standard frame as ($\mathbf{V}$), postponing the mention of localized vectors. Then ($\mathbf{V}$) may be associated legitimately with the origin (O) as base-point, and any element that might correspond to ($\mathbf{r}_0$) will be suppressed. With the usual unit-vectors, here taken at a common origin for convenience, we must have at the epoch, whatever range in orientation may be permitted for ($\mathbf{i}'\mathbf{j}'\mathbf{k}'$),

$$\mathbf{V} = \mathbf{i}V_{(x)} + \mathbf{j}V_{(y)} + \mathbf{k}V_{(z)} = \mathbf{i}'V_{(x')} + \mathbf{j}'V_{(y')} + \mathbf{k}'V_{(z')}. \quad (134)$$

This relation, to repeat with emphasis an incidental remark of section 79, may face in either of two directions, according as the data make ($\mathbf{V}$) itself or its three constituents directly known. The next equation derives much of its importance from the fact that the algebra so seldom furnishes a resultant vector immediately, unless the superficial geometry happens to fit.

Express now the time-derivative of (**V**); it will be consistently specified for the same standard frame as (**V**) itself, and it appears as

$$\dot{\mathbf{V}} = \mathbf{i}\frac{d}{dt}(V_{(x)}) + \mathbf{j}\frac{d}{dt}(V_{(y)}) + \mathbf{k}\frac{d}{dt}(V_{(z)})$$

$$= (\dot{\mathbf{i}}'V_{(x')} + \dot{\mathbf{j}}'V_{(y')} + \dot{\mathbf{k}}'V_{(z')})$$

$$+ \mathbf{i}'\frac{d}{dt}(V_{(x')}) + \mathbf{j}'\frac{d}{dt}(V_{(y')}) + \mathbf{k}'\frac{d}{dt}(V_{(z')}). \quad (135)$$

It is to be observed about tensors like $(V_{(x')})$ that they are differentiated on that comprehensive understanding about them, spoken of in section 78, which is favored by an algebra that attends to magnitudes alone and can neglect orientation. In the first group of the third member in this equation, it is the vector algebra with its equal attention to directions which is repairing that deficiency in the other algebra. In order to follow up and express this idea, we adopt the notation for all such cases,

$$\dot{\mathbf{V}}_{(m)} \equiv \mathbf{i}'\frac{d}{dt}(V_{(x')}) + \mathbf{j}'\frac{d}{dt}(V_{(y')}) + \mathbf{k}'\frac{d}{dt}(V_{(z')}), \quad (136)$$

intended to suggest that only the tensor magnitudes of (**i'j'k'**) have been differentiated. Omitting the second member of equation (135), and in reliance upon section 80 for a reduction of the first group, the third member can be rewritten in the more nearly standard form,

$$\dot{\mathbf{V}} = (\dot{\gamma} \times \mathbf{V}) + \dot{\mathbf{V}}_{(m)}. \quad (137)$$

But equation (134) would not be modified if the origin for (**i'j'k'**) were at any distance ($\mathbf{r}_0$) from (O) and were moving in any way. Our last result would still hold, provided the same ($\dot{\gamma}$) were retained, because it is a sheer relation for projections upon which it stands. Further, whenever ($\dot{\gamma}$) is zero, both (**V**) and (**V**) are represented indifferently by their respective components in (XYZ) or in (X′Y′Z′); and this harmonizes with the invariance found by using the permanent configurations of the coincidences and the idea of shift. Otherwise even when (**i′j′k′**) fall in (**ijk**) and make the two sets of components for (**V**) the same, the total time-derivatives of any algebraic expressions for the tensors of (**i′j′k′**) would not agree with the projections of ($\dot{\mathbf{V}}$) on (X′Y′Z′). But note that the proper partial derivatives of those tensors would give correct values for ($\dot{\mathbf{V}}$), as we discovered from equation (123) in the case of (**v**).

There is one condition of special arrangement that cancels the difference between ($\dot{\mathbf{V}}$) and ($\dot{\mathbf{V}}_{(m)}$) though ($\dot{\gamma}$) is not zero; namely, colinear or parallel factors in the corrective vector product. And since ($\dot{\gamma}$) as applying to (**i′j′k′**) rests on a supposition subject to a certain control, there is a strong hint in the above possibility of cancellation, which several coördinate systems have found their own ways to adopt. We can give a first illustration from our original discussion of the rotation-vector. For if we multiply equation (137) by (dt) and identify (**V**) with (dγ) the two members show equality to the second order, in confirmation of section 47.

86. Let the vector (**V**) be represented graphically from (O) as a base-point, in the manner of the velocity vector for the hodograph, then the derivative ($\dot{\mathbf{V}}$) will be given as the velocity of its extremity in (O, XYZ); and on comparing equations (111, 137), the former in application to a common origin, the other derivative ($\dot{\mathbf{V}}_{(m)}$) is seen to give similarly the velocity with which the extremity of (**V**) moves in the frame (X′Y′Z′). Consequently we find forms like ($\dot{\mathbf{V}}_{(m)}$) described sometimes as *derivatives relatively to the moving axes* (X′Y′Z′), and, to be sure, they are. But we must not neglect the other fact that this uncompleted derivative is applied to a quantity that like (**V**) has been specified for the standard frame, and that itself does not stand in any one particular relation to the frame (X′Y′Z′). These schemes, if thus viewed, are composite; or they straddle between the standard frame for (V) and a comparison-frame for ($\dot{\mathbf{V}}_{(m)}$); but they are less disjointed if interpreted as shift. The above denial, of course, runs only against a general truth, and does not exclude special conditions under which the same term covers both a shift and the other form of transfer. It is plain for example, in giving velocity by means of polar coördinates in uniplanar motion as

$$\mathbf{v} = \mathbf{r}_1 \frac{dr}{dt} + (\omega \times \mathbf{r}), \tag{138}$$

that the first term in the sum can be read either as ($\dot{\mathbf{V}}_{(m)}$), or as ($\mathbf{v}'$) for the frame consisting of (**r**) and a perpendicular, with the second term equally adapted to either sense.

It contributes much to the serviceable simplicity of equation (137) that it observes always the limits of a one-step transition from a vector to its first derivative, while a radical change of reference-frame must rebuild from the beginning by as many steps as are necessary. Let us exemplify how contrasts appear, by taking (**v**) as the vector of equation (137) and placing the result alongside equation (112), from which ($\dot{\mathbf{v}}_0$) has been removed by the supposition of a common origin, and in which, for closer parallelism, we have substituted for ($\mathbf{v}'$) in terms of (**v**). On one hand we find

$$\dot{\mathbf{v}} = (\dot{\gamma} \times \mathbf{v}) + \dot{\mathbf{v}}_{(m)}; \tag{139}$$

and on the other

$$\dot{\mathbf{v}} = (\ddot{\gamma} \times \mathbf{r}') + 2(\dot{\gamma} \times \mathbf{v}) - (\dot{\gamma} \times (\dot{\gamma} \times \mathbf{r}')) + \dot{\mathbf{v}}'. \tag{140}$$

It is evident how the latter equation has accumulated complications in its two steps that we followed earlier, and that the last terms in the two equations are not reduced to equality even by making ($\dot{\gamma}$) constant.

87. With this exposition accomplished, of the consequences for free vectors and their first derivatives of their inclusion in plans of shift, we can proceed to add for localized vectors those supplementary particulars which the localizing factor makes necessary in relations like

$$(\mathbf{r} \times \mathbf{V}) = (\mathbf{r}_0 \times \mathbf{V}) + (\mathbf{r}' \times \mathbf{V}), \tag{141}$$

when account is taken of the change in $(\mathbf{r}_0)$ due to shift of the comparison-frame into some new but permanent configuration. This allowance is obviously required in order to complete the details for the effective momentary replacement of (O, XYZ) by successive members in the group (O′, X′Y′Z′). And it is most easily disentangled from other elements, by using that superposition applying to similar cases which was indicated as far back as section 67.

Using the temporary notation

$$\mathbf{M} \equiv (\mathbf{r} \times \mathbf{V}); \qquad \mathbf{M}' \equiv (\mathbf{r}' \times \mathbf{V}); \tag{142}$$

the special question that concerns us here is the relation between $(\dot{\mathbf{M}})$ in the standard frame and $(\dot{\mathbf{M}}')$, the latter quantity being expressed under the guidance of ideas that it will be well to make quite explicit. First, the vector $(\mathbf{V})$ enters both products invariantly; and secondly, its total time-derivative appears without distinction in both, because changes in $(\mathbf{i}'\mathbf{j}'\mathbf{k}')$ being now put aside in order to consider changes in $(\mathbf{r}_0)$ alone, the corrective term of equation (137) disappears. But thirdly, with (γ) dropped from the list of section 78 for the reason named, $(\mathbf{r}')$ becomes a function of the two variables $(\mathbf{r}_0, t)$. Then its exact differential is for the present shift

$$d\mathbf{r}' = \frac{\partial \mathbf{r}'}{\partial \mathbf{r}_0} d\mathbf{r}_0 + \frac{\partial \mathbf{r}'}{\partial t} dt; \tag{143}$$

and if this is timed to march with the actual changes during (dt) we get

$$\frac{d\mathbf{r}'}{dt} = \frac{\partial \mathbf{r}'}{\partial \mathbf{r}_0}\frac{d\mathbf{r}_0}{dt} + \frac{\partial \mathbf{r}'}{\partial t}; \qquad \frac{d\mathbf{r}_0}{dt} \equiv \dot{\mathbf{r}}_0; \qquad \frac{\partial \mathbf{r}'}{\partial t} = \dot{\mathbf{r}}; \tag{144}$$

the last equality having the same validity as in equation (124). Hence

$$\dot{\mathbf{M}}' = \left(\frac{\mathrm{d}\mathbf{r}'}{\mathrm{dt}} \times \mathbf{V}\right) + (\mathbf{r}' \times \dot{\mathbf{V}}) = ((\dot{\mathbf{r}} - \dot{\mathbf{r}}_0) \times \mathbf{V}) + (\mathbf{r}' \times \dot{\mathbf{V}}); \quad (145)$$

$$\dot{\mathbf{M}} = (\dot{\mathbf{r}} \times \mathbf{V}) + (\mathbf{r} \times \dot{\mathbf{V}}) = \dot{\mathbf{M}}' + (\dot{\mathbf{r}}_0 \times \mathbf{V}) + (\mathbf{r}_0 \times \dot{\mathbf{V}}). \quad (146)$$

Consequently, though (O′) coincides with (O), if there is displacement of the former with shift rate ($\dot{\mathbf{r}}_0$) the values of ($\dot{\mathbf{M}}$) and ($\dot{\mathbf{M}}'$) as defined will still differ by the term ($\dot{\mathbf{r}}_0 \times \mathbf{V}$).

We may restate the last equation by arriving at it through

$$\mathbf{M} - \mathbf{M}' = (\mathbf{r}_0 \times \mathbf{V}); \qquad \dot{\mathbf{M}} - \dot{\mathbf{M}}' = (\dot{\mathbf{r}}_0 \times \mathbf{V}) + (\mathbf{r}_0 \times \dot{\mathbf{V}}), \quad (147)$$

if that is deemed a sufficient analysis of the conditions for the differentiation; and there is precedent for calling ($\dot{\mathbf{M}}'$) the moment of ($\dot{\mathbf{V}}$) for a moving base-point. It is only iteration here, however, to make the comment that the directer thought holds in view the stationary points (O′), for which the coincident moving point serves as marker at beginning and end of the interval (dt).

Let us make application of this development to moment of momentum and its derivative, as being the localized vectors among our fundamental quantities. We are still confining attention to shift of origin alone; and we shall not go beyond the expressions for the representative particle at the center of mass. Write then

$$\mathbf{H} = (\bar{\mathbf{r}} \times \mathbf{Q}) = (\mathbf{r}_0 + \bar{\mathbf{r}}') \times \mathbf{Q} = (\mathbf{r}_0 \times \mathbf{Q}) + \mathbf{H}_{(o')}; \quad (148)$$

$$\dot{\mathbf{M}}' = ((\bar{\mathbf{v}} - \dot{\mathbf{r}}_0) \times \mathbf{Q}) + (\bar{\mathbf{r}}' \times \dot{\mathbf{Q}}) = \dot{\mathbf{H}}_{(o')};$$

and reduce by omitting the product of colinear factors. But for the moment about (O′) of the force measured in the standard frame we have

$$\mathbf{M}(_{o'}) \equiv \bar{\mathbf{r}}' \times \dot{\mathbf{Q}} = \dot{\mathbf{H}}(_{o'}) + (\dot{\mathbf{r}}_0 \times \mathbf{Q}), \quad (149)$$

which thus replaces with these conditions of shift the relation of equation (VI).

88. For establishing the theorem of equation (137) and presenting its bearings and a few of its consequences, reliance has been placed almost exclusively upon the vector algebra; yet those ideas were manageable to the other algebra also, though it cannot fail to be apparent how much the absence there of direct indication for orientation renders the operations in matters like these more cumbrous, and the expressed results less perspicuous. If, therefore, it seems profitable to go over part of that ground in terms of the older method, that is not at all with wasted effort upon verification, nor in order to gain reward in fuller insight, except as seeing the cross connections is likely to prove instructive. But coördinate algebra is indispensable for calculation; transition to more succinct treatment, where it can finally displace the older method, is still in progress, which is keeping some comparisons temporarily that will fall away later; and moreover, the next chapter is concerned with coördinate systems as its chief topic. Consequently in preparation for that material and for these other reasons, it seems well to put in a link of connection; we shall, therefore, proceed to parallel section 85 with the algebraic equations that offer the same meaning under other forms.

It is unnecessary to carry a separation of origins into this development, because as we have noticed repeatedly its effects are in themselves easy to record, and are cared for completely by uncomplicated superposition. Thinking of (X′Y′Z′) and (XYZ) as having common origin (O), (x′y′z′) and (xyz) are, in the first instance, the coördinates of any point (Q). But we can draw advantage in two ways from previous experience; first, (Q) can locate a representative particle of finite mass as well as one mass-element of a body, and secondly, (**x′y′z′**) and (**xyz**) can be made to denote the projections of any vector (**V**) with base-point at (O), by extension of their relation to the particular vector (**r**) that is now identical with (**r**′). Unless the contrary is said explicitly, (**V**) is to be regarded as determined in the standard frame (XYZ), and introduced invariantly into any connections with (X′Y′Z′). This vector can be regarded as localized at (O) either by its property as a recognized free vector like (**Q**) and (**R**), or by a convention agreeing with its nature in cases like the rotation-vector (ω) and its companions ($\dot{\omega}$), (**H**), and (**M**) when pure rotation about (O) is supposed. The symbols are to be endowed with the wider valid meanings in the equations constructed according to the adjoining table that shows the direction cosines of the relative configuration.

X′	Y′	Z′	with
l_1	m_1	n_1	X
l_2	m_2	n_2	Y
l_3	m_3	n_3	Z

89. The usual transformation equations when made explicit for (xyz) are

$$\left.\begin{aligned} x &= l_1x' + m_1y' + n_1z', \\ y &= l_2x' + m_2y' + n_2z', \\ z &= l_3x' + m_3y' + n_3z'. \end{aligned}\right\} \qquad (150)$$

And the companion forms derivable by an elementary process are

$$\left.\begin{aligned} x' &= l_1x + l_2y + l_3z, \\ y' &= m_1x + m_2y + m_3z, \\ z' &= n_1x + n_2y + n_3z. \end{aligned}\right\} \qquad (151)$$

Together these are known to depend upon or to express the mutual relations of projection between two sets of components of the same resultant vector. When the direction cosines are invariable, the correspondence with constancy of (**i**′**j**′**k**′) is evident, and the same mutual relation runs on into all the derivatives, giving invariance whose obvious details need not detain us. A change of configuration, however, makes in general all the direction cosines vary, and there the same alternatives recur that were brought out in sections 78 and 82. One of these will make (x′, y′, z′) each a function of three independent variables that are time and two direction cosines, the third of the latter being removed by a standard connection like

$$l_1^2 + l_2^2 + l_3^2 = 1. \tag{152}$$

The second point of view will set time in its place as the one independent variable of which all other quantities are functions; but here it will be just as desirable as before to put into properly conspicuous relief the modified relation of time to variables like (x, y, z) and to others like (l_1, l_2, l_3).

90. Equations of the same type as

$$\frac{\partial x'}{\partial t} = l_1 \frac{dx}{dt} + l_2 \frac{dy}{dt} + l_3 \frac{dz}{dt} \tag{153}$$

can be read in the light of equation (123); and what remain to examine are the complete time-derivatives of the quantities (x′y′z′), principally in order to detect the rotation-vector ($\dot{\gamma}$) of (X′Y′Z′) by penetrating its disguise of direction angles and their derivatives. Adopting the fluxion notation, for ease in writing total time-derivatives, we have first

$$\dot{x}' = (l_1\dot{x} + l_2\dot{y} + l_3\dot{z}) + (\dot{l}_1 x + \dot{l}_2 y + \dot{l}_3 z). \tag{154a}$$

Note in passing, as consequences of equations (151, 154) that may prove suggestive later,[1]

$$l_1 = \frac{\partial x'}{\partial x} = \frac{\partial \dot{x}'}{\partial \dot{x}}; \qquad \dot{l}_1 = \frac{d}{dt}\left(\frac{\partial x'}{\partial x}\right) = \frac{\partial \dot{x}'}{\partial x}; \tag{155}$$

which are typical of similar relations running all through the

[1] See Note 23.

sets of equations, when we add to the value of $(\dot{x}')$ its companions

$$\left.\begin{aligned} \dot{y}' &= (m_1\dot{x} + m_2\dot{y} + m_3\dot{z}) + (\dot{m}_1 x + \dot{m}_2 y + \dot{m}_3 z), \\ \dot{z}' &= (n_1\dot{x} + n_2\dot{y} + n_3\dot{z}) + (\dot{n}_1 x + \dot{n}_2 y + \dot{n}_3 z). \end{aligned}\right\} \quad (154b)$$

Concentrating attention upon the last groups in these equations, because the effects of changing configuration appear exclusively in them, and introducing the necessary direction angles in order to prepare for the connection with $(\dot{\gamma})$, expand into the forms

$$\left.\begin{aligned} &- [x\dot{\alpha}_1 \sin \alpha_1 + y\dot{\alpha}_2 \sin \alpha_2 + z\dot{\alpha}_3 \sin \alpha_3]; \\ &- [x\dot{\beta}_1 \sin \beta_1 + y\dot{\beta}_2 \sin \beta_2 + z\dot{\beta}_3 \sin \beta_3]; \\ &- [x\dot{\epsilon}_1 \sin \epsilon_1 + y\dot{\epsilon}_2 \sin \epsilon_2 + z\dot{\epsilon}_3 \sin \epsilon_3]. \end{aligned}\right\} \quad (156)$$

But the normal to the plane (X′, X) must be the axis for $(\dot{\alpha}_1)$; and with the direction cosines of those intersecting lines given as

$$1,\ 0,\ 0,\ (X); \qquad l_1,\ l_2,\ l_3,\ (X'); \qquad (157)$$

the direction cosines (λ, μ, ν) of the normal to their plane worked out by the standard method gives

$$\lambda = 0; \qquad \mu = -\frac{\cos \alpha_3}{\sin \alpha_1}; \qquad \nu = \frac{\cos \alpha_2}{\sin \alpha_1}. \qquad (158)$$

But as explained in section 46 the rate at which (X′) is turning about that normal must be the projection of $(\dot{\gamma})$ upon that line, or equivalently,

$$\dot{\alpha}_1 = \lambda\dot{\gamma}_{(x)} + \mu\dot{\gamma}_{(y)} + \nu\dot{\gamma}_{(z)}, \qquad (159)$$

from which follows

$$-\dot{\alpha}_1 \sin \alpha_1 = \dot{\gamma}_{(y)} \cos \alpha_3 - \dot{\gamma}_{(z)} \cos \alpha_2. \qquad (160)$$

Proceeding similarly with the eight other terms which complete the group of that type in equations (154), it is seen after simple reduction that they make up in the first, second and third equation respectively

$$\left.\begin{aligned} &-(\dot{\gamma}_{(y')}z' - \dot{\gamma}_{(z')}y'); \qquad -(\dot{\gamma}_{(z')}x' - \dot{\gamma}_{(x')}z' \\ &\qquad -(\dot{\gamma}_{(x')}y' - \dot{\gamma}_{(y')}x'). \end{aligned}\right. \qquad (161)$$

Since the first members of those equations correspond to the total derivatives of the tensors obtainable from equation (125), we find after orientation and forming the vector sum that equations (154) yield consistently with equation (137)

$$\dot{\mathbf{V}}_{(m)} = \dot{\mathbf{V}} - (\dot{\gamma} \times \mathbf{V}), \tag{162}$$

on our understanding about the broader meaning of ($\mathbf{x'y'z'}$) and ($\mathbf{xyz}$).

It is left as an exercise, modeled on the above plan but continued into the formation of second derivatives, to reach by the algebraic routine the coördinate equations which together represent the result recorded in equation (112), if we suppress there all terms depending on a separation of origins. Where the quantity ($\ddot{\gamma}$) occurs in executing this, it is of interest to realize what has been alluded to elsewhere; that ($\dot{\gamma}$) and ($\ddot{\gamma}$) may be connected with either (XYZ) or (X′Y′Z′), since the difference term in equation (162) is zero when ($\dot{\gamma}$) is ($\mathbf{V}$).

CHAPTER IV

The Main Coördinate Systems

91. The standard frame itself has an additional office of providing a coördinate system that is basic in certain ways, and that is in fact tacitly utilized for the semi-cartesian expansions in terms of (**ijk**), both in immediate relation to vector quantities, and for the expression of constituents in work, kinetic energy and power, where vector factors occur in scalar products. To do these things has become so much habitual or even instinctive that we learn with some surprise how Maclaurin is given credit for invention here, as Euler is for inventing the concept of fluid pressure, which at this date might also seem part of external nature.

The standard frame, too, has one lead in advantage over other resolutions through the unqualified permanence of its origin and of its unit-vectors, which enables us to submit its tensors unhesitatingly to algebraic operations, and pass over to vector algebra by merely supplying the ellipsis of the unaffected orienting factors. The disturbing influences in other combinations, where ($\mathbf{r}_0$) and (**i′j′k′**) make more caution advisable, have been forcing themselves upon us repeatedly. But as we have seen illustrated for mean values, and as is not unusual, the presence of such desirable elements as we find in the standard frame may be also a drawback. Within the complete projection on a standard axis, distinctions of source in changes of magnitude or of direction may be lost, that are vital in the vectors that play a part. The net force parallel to (X) and its work, if written for a particle

$$X = m\frac{d^2x}{dt^2}\,; \qquad W = \int X dx; \tag{163}$$

hide, in the first, the fact that normal force (**N**) and tangential force (**T**) are coalescing in the one sum, and in the second, that part of this work is illusory in so far as the projection of (**N**) enters the sum (X), and does work in the algebra though not in the mechanics. At one other point we have been enabled to compare the principal axes of inertia with (XYZ) and ascertain that all advantage does not lie with the latter, for expressing compactly either the scalar energy or the vector force-moment. And these considerations, in sum, may justify us in leaving the resolution into constituents according to the standard axes to one side, except where we touch upon it for some special connection. Then we are free to devote detailed attention to other coördinate systems that are chiefly current, and make due analysis of their intention and of the scope of their success.

It seems quite enough therefore if we collect here the indicated partitions for (XYZ) that are reasonably self-evident rewritings of the totals to which the preceding text has given most weight:

$$\mathbf{Q} = \mathbf{i}\Sigma\int_m \dot{x}dm + \mathbf{j}\Sigma\int_m \dot{y}dm + \mathbf{k}\Sigma\int_m \dot{z}dm; \tag{164}$$

$$\mathbf{H} = \mathbf{i}\Sigma\int_m (y\dot{z} - z\dot{y})dm + \mathbf{j}\Sigma\int_m (z\dot{x} - x\dot{z})dm + \mathbf{k}\Sigma\int_m (x\dot{y} - y\dot{x})dm; \tag{165}$$

$$E = \tfrac{1}{2}\Sigma\int_m \dot{x}^2dm + \tfrac{1}{2}\Sigma\int_m \dot{y}^2dm + \tfrac{1}{2}\Sigma\int_m \dot{z}^2dm; \tag{166}$$

$$\mathbf{R} = \mathbf{i}\Sigma\int_m \ddot{x}dm + \mathbf{j}\Sigma\int_m \ddot{y}dm + \mathbf{k}\Sigma\int_m \ddot{z}dm; \tag{167}$$

$$\mathbf{M} = \mathbf{i}\Sigma\int_m (y\ddot{z} - z\ddot{y})dm + \mathbf{j}\Sigma\int_m (z\ddot{x} - x\ddot{z})dm + \mathbf{k}\Sigma\int_m (x\ddot{y} - y\ddot{x})dm; \tag{168}$$

$$P = \Sigma\int_m \dot{x}dX + \Sigma\int_m \dot{y}dY + \Sigma\int_m \dot{z}dZ. \tag{169}$$

It will be found profitable to compare equations (165) and (86); also equations (166) and (81, 88), including the comment preceding the latter. Since the first three equations in the above group are mere expansions of the forms in section 15, they have the same scope as those. Similarly the validity of the last three is coextensive with that for equations (16, 17, 18) of which they are the expansions.

EULER'S CONFIGURATION ANGLES.

92. Because it deals directly and exclusively with the recurrent element that is found at the root of so many particular results, we shall take up next those orientation angles for specifying configuration which were devised by Euler and by custom bear his name. They have not yet been displaced from a conceded position of value in use for their purpose. There is an added reason for giving these angles proper discussion in that the expression of them as vectors has scarcely been attempted; we find their connections with other specifying elements almost exclusively in the form of purely algebraic equations. It is a curious fact that angle in prevailing practice has not arrived at legal recognition as a vector, though the vector quality of its first and second time-derivatives, angular velocity and angular acceleration, was announced and employed a number of years ago. So we need to do something consciously toward incorporating angle-vectors into our scheme of treatment on a parity with other vector quantities, in order that real symmetries of relation may not be seen distorted.

Supposing that one end of a line (**r**) is fixed and that it moves into a new position, its second configuration in relation to its first can be given by a vector-angle normal to the plane of the two positions. This vector is axial, and related to an area with duly assigned circulation; and the area is in the plane located by the extreme positions of (**r**), its magnitude being twice that of the sector of the unit circle limited by those positions. But such a direct representation of this total would be no more convenient for use in all cases than other resultants are, so its projections according to Euler's plan are substituted, which amounts to giving the latitude and the longitude on unit sphere centered at the fixed point or origin (O), in which (r) cuts that surface. Assuming next that (**r**) is a definite line of a rigid solid that is limited to pure rotation about (O), a third angle added will enable us to complete the description of a new configuration for the solid, and this last angle will denote a rotational displacement about (**r**). We shall follow usage in assigning the symbols ($\boldsymbol{\vartheta}$) to the *latitude angle*, and ($\boldsymbol{\psi}$) to the *longitude angle*, while ($\boldsymbol{\varphi}$) is added for the rotation about (**r**); it remains only to agree upon zero values of the three angular coördinates. It suits our purpose in its general course better, to think in terms of a displaced *rigid cross* (X′Y′Z′), which may here be made equivalent to the rigid solid named above, and then coincidence of (X′Y′Z′) with (XYZ) yields the natural zero. We identify (Z) with the earth's polar axis in its relation to latitude and longitude.

93. Beginning with resultant angular displacement ($\boldsymbol{\gamma}$) at zero, and (X′Y′Z′) coincident with (XYZ), let the plane (Y′Z′) separate from (YZ) by angular displacement ($\boldsymbol{\psi}$) about (Z), in which that vector angle must then fall. Next let angular displacement ($\boldsymbol{\vartheta}$) occur about the displaced position of (X′), in whose line therefore it must lie as a vector angle; and finally let (X′, Y′) turn with angular displacement ($\boldsymbol{\varphi}$) about the final position of (Z′), with whose line this third vector angle must then coincide. To make the conditions standard, ($\boldsymbol{\psi}$, $\boldsymbol{\vartheta}$, $\boldsymbol{\varphi}$) are all to be taken positive by the rule of the right-handed cycle. The order of the three displacements has been chosen so that each is made about one of the three axes (X′Y′Z′) as found at the beginning of that stage. It is verified without difficulty that the summed projections on (XYZ) are

$$\left.\begin{aligned} \boldsymbol{\gamma}_{(x)} &= \mathbf{i}(\vartheta \cos \psi + \varphi \sin \vartheta \sin \psi); \\ \boldsymbol{\gamma}_{(y)} &= \mathbf{j}(\vartheta \sin \psi - \varphi \sin \vartheta \cos \psi); \\ \boldsymbol{\gamma}_{(z)} &= \mathbf{k}(\psi + \varphi \cos \vartheta). \end{aligned}\right\} \qquad (170)$$

And if we resolve on the final orientations of (X′Y′Z′), those projections are

$$\left.\begin{aligned}\gamma_{(x')} &= \mathbf{i}'(\vartheta \cos \varphi + \psi \sin \vartheta \sin \varphi);\\ \gamma_{(y')} &= \mathbf{j}'(-\vartheta \sin \varphi + \psi \sin \vartheta \cos \varphi);\\ \gamma_{(z')} &= \mathbf{k}'(\varphi + \psi \cos \vartheta).\end{aligned}\right\} \qquad (171)$$

These two sets of projections are orthogonal; but if we state the supposed displacements directly, and let ($\boldsymbol{\psi}_1$, $\boldsymbol{\vartheta}_1$, $\boldsymbol{\varphi}_1$) represent unit-vectors agreeing with those suppositions, the set is oblique to the extent that the angle ($\boldsymbol{\psi}_1$, $\boldsymbol{\varphi}_1$) is ($\boldsymbol{\vartheta}$) and not in general a right angle. We add accordingly,

$$\gamma = \boldsymbol{\psi}_1(\psi) + \boldsymbol{\vartheta}_1(\vartheta) + \boldsymbol{\varphi}_1(\varphi), \qquad (172)$$

and have secured three equivalent forms of expression for the resultant angle-vector (γ). Observe also the differences among the three in regard to the unit-vectors; (**ijk**) are permanently oriented, (**i′j′k′**) are capable of displacement by rotation, for they remain orthogonal, but ($\boldsymbol{\psi}_1$, $\boldsymbol{\vartheta}_1$, $\boldsymbol{\varphi}_1$) must be considered individually. It is seen, if we hold definitely to the terms of the description, that ($\boldsymbol{\psi}_1$) is of permanent orientation in (Z), that ($\boldsymbol{\vartheta}_1$) depends for orientation upon ($\boldsymbol{\psi}$), being always normal to the displaced position of the (Y′Z′) plane, and that ($\boldsymbol{\varphi}_1$) depends similarly upon both ($\boldsymbol{\psi}$) and ($\boldsymbol{\vartheta}$), because the ($\boldsymbol{\varphi}$) displacement begins where the second stage leaves off. All three quantities ($\boldsymbol{\psi}$, $\boldsymbol{\vartheta}$, $\boldsymbol{\varphi}$) are rotation-vectors applying to the axis-set (X′Y′Z′) as representative of a rigid body, and standing to the changes of direction of individual lines in the relation established by section 46. This needs to be borne in mind if any question should be opened about changing the sequence of the three steps, so that ($\boldsymbol{\vartheta}$) and ($\boldsymbol{\varphi}$) though equal to their first magnitudes are connected as vectors with different axes.

The above forms of statement are mathematically on the same footing as a means of determining (γ), but there can be no real doubt where the preference would fall on the score of ease in application or execution, when the three plans are compared. The second is especially intricate because its projections are associated with that very terminal configuration of (X′Y′Z′) which it may be the object to locate, but which must somehow become known before the scheme can assume full definiteness. It should be inserted however for the sake of its subsequent uses.

94. The employment of the standard angles (ψ, ϑ, ϕ) is not confined to expressing configurations, and is therefore not exhausted in equations (170, 171, 172). Indeed the primary service of Euler's so-called *geometrical equations* has begun at their developed connections with the rotation-vector or angular velocity, and found a natural continuation in dealing with angular acceleration written ($\ddot{\gamma}$) or ($\dot{\omega}$). As we now undertake to make those connections clear, combinations will occur at first or in later application, that make it advisable to retain ($\dot{\gamma}$) and ($\ddot{\gamma}$) for use with comparison-frames like (X′Y′Z′), and let the meaning of the parallel quantities (ω) and ($\dot{\omega}$) refer exclusively, as in sections 45, 55, 62 and 63, to a rigid body's rotation, either about its center of mass or about some fixed point. To maintain this consistent distinction will avoid confusion where both pairs of elements are presented in the same inquiry.

The expressions for (γ) that we have just obtained are contrived to show its value at the advancing front of a progressive angular displacement to which (ψ, ϑ, ϕ) can be considered to belong. Consequently it is adapted to differentiation, with a view to exhibit either a systematic succession of partial differentials or simultaneous time rates in a total derivative; and previous discussions have laid a foundation for interpretations leading in both directions. In the first instance we are most nearly concerned with the derivation of ($\dot{\gamma}$) from the three several equations (170, 171, 172) and the collation of results with section 85 as bearing upon the current algebraic forms. And because this has some little flavor of revising the latter, the fuller infusion of vector peculiarities into these matters having not yet worn off its novelty, there seems to exist a stronger reason for detail, than the mere arrival at conclusions for handy use might demand.[1]

95. As in similar comparisons elsewhere, the (**ijk**) projections furnish reliably through pure and total tensor differentiation an unquestioned standard to which alternatives must conform if correctly formulated. So the first straightforward step is to employ equation (170) in this test; and we prepare the way with the expansion

$$\begin{aligned}\dot{\gamma} = \mathbf{i}\Bigg[&\left(\frac{d\vartheta}{dt}\cos\psi + \frac{d\varphi}{dt}\sin\vartheta\sin\psi\right)\\ &+\left(-\vartheta\sin\psi\frac{d\psi}{dt} + \varphi\cos\vartheta\sin\psi\frac{d\vartheta}{dt}\right.\\ &\qquad\left. + \varphi\sin\vartheta\cos\psi\frac{d\psi}{dt}\right)\Bigg]\\ +\mathbf{j}\Bigg[&\left(\frac{d\vartheta}{dt}\sin\psi - \frac{d\varphi}{dt}\sin\vartheta\cos\psi\right)\\ &+\left(\vartheta\cos\psi\frac{d\psi}{dt} - \varphi\cos\vartheta\cos\psi\frac{d\vartheta}{dt}\right.\\ &\qquad\left. + \varphi\sin\vartheta\sin\psi\frac{d\psi}{dt}\right)\Bigg]\\ +\mathbf{k}\Bigg[&\left(\frac{d\psi}{dt} + \frac{d\varphi}{dt}\cos\vartheta\right) + \left(-\varphi\sin\vartheta\frac{d\vartheta}{dt}\right)\Bigg].\end{aligned} \tag{173}$$

But we have been remarking from section 79 onward that the partial time-derivatives in equations like (171, 172), when the unit-vectors are made variables, must reproduce the standard frame values obtained through (**ijk**). Let us accordingly write

[1] See Note 24.

out those two sets of partials and proceed toward comparing them with equation (173). Observing that the conditions of the differentiation exclude trigonometric functions of the angles from varying, though they permit the angles as magnitudes to change, we find

$$\left.\begin{aligned}\frac{\partial \Upsilon}{\partial t}(\mathrm{i}', \mathrm{j}', \mathrm{k}') = \mathbf{i}'\left(\frac{\partial \vartheta}{\partial t}\cos\varphi + \frac{\partial \psi}{\partial t}\sin\vartheta\sin\varphi\right) \\ + \mathbf{j}'\left(-\frac{\partial \vartheta}{\partial t}\sin\varphi + \frac{\partial \psi}{\partial t}\sin\vartheta\cos\varphi\right) \\ + \mathbf{k}'\left(\frac{\partial \varphi}{\partial t} + \frac{\partial \psi}{\partial t}\cos\vartheta\right),\end{aligned}\right\} \quad (174)$$

$$\frac{\partial \Upsilon}{\partial t}(\psi_1, \vartheta_1, \varphi_1) = \boldsymbol{\psi}_1\left(\frac{\partial \psi}{\partial t}\right) + \boldsymbol{\vartheta}_1\left(\frac{\partial \vartheta}{\partial t}\right) + \boldsymbol{\varphi}_1\left(\frac{\partial \varphi}{\partial t}\right). \quad (175)$$

The value directly apparent in the last equation can be noticed by inspection to agree with that of the equation preceding, if we assemble mentally from the latter the items falling respectively along $(\boldsymbol{\psi}_1, \boldsymbol{\vartheta}_1, \boldsymbol{\varphi}_1)$. And this coincidence is next to be recognized similarly in the first groups marked off under (**i**, **j**, **k**) in equation (173), with the single variation that the latter appear as total derivatives of the angle magnitudes. The patent conclusion is that proper allowance for the difference between these total and these partial time-derivatives must exactly offset the remaining groups in equation (173); and that outcome might be accepted on the fair ground that it harmonizes with equations (126, 131), without going further. Yet the completed analysis of how that compensation is in fact brought about here, has an immediate bearing and interest that justify setting down its several steps.

96. The last groups of terms in equation (173) can be brought together and rearranged so that they are identified as the vector products to which they are equated below:

$$\begin{aligned}\vartheta \frac{d\psi}{dt} & (-\mathbf{i} \sin \psi + \mathbf{j} \cos \psi) \\ & + \varphi \frac{d\vartheta}{dt} (\mathbf{i} \cos \vartheta \sin \psi - \mathbf{j} \cos \vartheta \cos \psi - \mathbf{k} \sin \vartheta) \\ & + \varphi \frac{d\psi}{dt} (\mathbf{i} \sin \vartheta \cos \psi + \mathbf{j} \sin \vartheta \sin \psi) \\ & = \frac{d\psi}{dt} \vartheta(\boldsymbol{\psi}_1 \times \boldsymbol{\vartheta}_1) + \frac{d\vartheta}{dt} \varphi(\boldsymbol{\vartheta}_1 \times \boldsymbol{\varphi}_1) + \frac{d\psi}{dt} \varphi(\boldsymbol{\psi}_1 \times \boldsymbol{\varphi}_1).\end{aligned} \tag{176}$$

The verification as regards magnitudes, directions and order of factors in the vector products is ordinary routine devoid of artifice, due regard being paid to the specifications of direction in the sections immediately preceding. The character of the second member is plain: it consists of allowances for changing directions of the two unit-vectors $(\boldsymbol{\vartheta}_1)$ and $(\boldsymbol{\varphi}_1)$, the former being affected by the turning about $(\boldsymbol{\psi}_1)$, and the latter by the two turnings about $(\boldsymbol{\psi}_1)$ and $(\boldsymbol{\vartheta}_1)$. It is instructive to notice that these individual consequences of the changes in the unit-vectors preserve their type and enter singly in parallel with the developments of sections 47 and 80, although there is here no common factor, the rotation-vector, related equally to all three unit-vectors $(\boldsymbol{\psi}_1, \boldsymbol{\vartheta}_1, \boldsymbol{\varphi}_1)$. This line of attack has been adopted partly in order to extend in that direction our earlier proof.

In preparing to demonstrate that the differences between $(\partial\vartheta/\partial t)$ and $(d\vartheta/dt)$, $(\partial\varphi/\partial t)$ and $(d\varphi/dt)$, exactly nullify the second member of equation (176), it is most direct to start from explicit values of $(\psi, \vartheta, \varphi)$. By a process of elementary elimination applied to equations (170) it follows that

$$\left.\begin{aligned}\psi &= \gamma_{(z)} - \frac{\cos \vartheta}{\sin \vartheta} (\gamma_{(x)} \sin \psi - \gamma_{(y)} \cos \psi); \\ \vartheta &= \gamma_{(x)} \cos \psi + \gamma_{(y)} \sin \psi; \\ \varphi &= \frac{1}{\sin \vartheta} (\gamma_{(x)} \sin \psi - \gamma_{(y)} \cos \psi).\end{aligned}\right\} \tag{177}$$

It is to be remarked as regards these equations that in order to arrive at their partial time-derivatives, we must include as variables only $(\gamma_{(x)})$, $(\gamma_{(y)})$, $(\gamma_{(z)})$, and for the total derivatives we must include also all the other factors as functions of time. It is therefore possible to write these indications of the differences:

$$\left.\begin{aligned}
\frac{d\psi}{dt} - \frac{\partial\psi}{\partial t} &= \frac{\partial\psi}{\partial\psi}\frac{d\psi}{dt} + \frac{\partial\psi}{\partial\vartheta}\frac{d\vartheta}{dt};\\
\frac{d\vartheta}{dt} - \frac{\partial\vartheta}{\partial t} &= \frac{\partial\vartheta}{\partial\psi}\frac{d\psi}{dt};\\
\frac{d\varphi}{dt} - \frac{\partial\varphi}{\partial t} &= \frac{\partial\varphi}{\partial\psi}\frac{d\psi}{dt} + \frac{\partial\varphi}{\partial\vartheta}\frac{d\vartheta}{dt}.
\end{aligned}\right\} \qquad (178)$$

Evaluating the second members from equations (177) and finally adding the orienting unit-vectors we derive these expressions:

$$\left.\begin{aligned}
\boldsymbol{\psi}_1\frac{\partial\psi}{\partial\psi}\frac{d\psi}{dt} &= -\boldsymbol{\psi}_1\frac{\cos\vartheta}{\sin\vartheta}(\gamma_{(x)}\cos\psi + \gamma_{(y)}\sin\psi)\frac{d\psi}{dt}\\
&= \boldsymbol{\psi}_1\left(-\frac{\cos\vartheta}{\sin\vartheta}\vartheta\frac{d\psi}{dt}\right);\\
\boldsymbol{\psi}_1\frac{\partial\psi}{\partial\vartheta}\frac{d\vartheta}{dt} &= \boldsymbol{\psi}_1\frac{1}{\sin^2\vartheta}(\gamma_{(x)}\sin\psi - \gamma_{(y)}\cos\psi)\frac{d\vartheta}{dt}\\
&= \boldsymbol{\psi}_1\left(\frac{1}{\sin\vartheta}\varphi\frac{d\vartheta}{dt}\right);\\
\boldsymbol{\vartheta}_1\frac{\partial\vartheta}{\partial\psi}\frac{d\psi}{dt} &= \boldsymbol{\vartheta}_1(-\gamma_{(x)}\sin\psi + \gamma_{(y)}\cos\psi)\frac{d\psi}{dt}\\
&= \boldsymbol{\vartheta}_1\left(-\varphi\sin\vartheta\frac{d\psi}{dt}\right);\\
\boldsymbol{\varphi}_1\frac{\partial\varphi}{\partial\psi}\frac{d\psi}{dt} &= \boldsymbol{\varphi}_1\frac{1}{\sin\vartheta}(\gamma_{(x)}\cos\psi + \gamma_{(y)}\sin\psi)\frac{d\psi}{dt}\\
&= \boldsymbol{\varphi}_1\left(\frac{1}{\sin\vartheta}\vartheta\frac{d\psi}{dt}\right);\\
\boldsymbol{\varphi}_1\frac{\partial\varphi}{\partial\vartheta}\frac{d\vartheta}{dt} &= \boldsymbol{\varphi}_1\left(-\frac{\cos\vartheta}{\sin^2\vartheta}(\gamma_{(x)}\sin\psi - \gamma_{(y)}\cos\psi)\right)\frac{d\vartheta}{dt}\\
&= \boldsymbol{\varphi}_1\left(-\frac{\cos\vartheta}{\sin\vartheta}\varphi\frac{d\vartheta}{dt}\right).
\end{aligned}\right\} \qquad (179)$$

After forming them into three groups as shown below, they can be recognized as constituting the vector products to which they are severally equated;

$$\left.\begin{aligned} \vartheta\frac{d\psi}{dt}\left(-\boldsymbol{\psi}_1\frac{\cos\vartheta}{\sin\vartheta}+\boldsymbol{\phi}_1\frac{1}{\sin\vartheta}\right) &= -\frac{d\psi}{dt}\vartheta(\boldsymbol{\psi}_1\times\boldsymbol{\vartheta}_1); \\ \varphi\frac{d\vartheta}{dt}\left(\boldsymbol{\psi}_1\frac{1}{\sin\vartheta}-\boldsymbol{\phi}_1\frac{\cos\vartheta}{\sin\vartheta}\right) &= -\frac{d\vartheta}{dt}\varphi(\boldsymbol{\vartheta}_1\times\boldsymbol{\phi}_1); \\ -\boldsymbol{\vartheta}_1\varphi\frac{d\psi}{dt}\sin\vartheta &= -\frac{d\psi}{dt}\varphi(\boldsymbol{\psi}_1\times\boldsymbol{\phi}_1). \end{aligned}\right\} \quad (180)$$

The first quantity of these three is known by the first parenthesis to be perpendicular to $(\boldsymbol{\psi}_1)$ in the plane of $(\boldsymbol{\psi}_1, \boldsymbol{\phi}_1)$; so the second quantity is perpendicular to $(\boldsymbol{\phi}_1)$ in the same plane; and $(\boldsymbol{\vartheta}_1)$ is by supposition normal to that plane. The directions match the order of factors and the signs.

97. When the established conclusions of equations (176, 180) are united with what was found to be true on casting up into a vector sum the three first groups in the coefficients of (**i**, **j**, **k**), equation (173), the registration of all these connections yields the continued equality

$$\begin{aligned} \dot{\boldsymbol{\gamma}} &= \mathbf{i}\left(\frac{\partial\vartheta}{\partial t}\cos\psi+\frac{\partial\varphi}{\partial t}\sin\vartheta\sin\psi\right) \\ &\quad+\mathbf{j}\left(\frac{\partial\vartheta}{\partial t}\sin\psi-\frac{\partial\varphi}{\partial t}\sin\vartheta\cos\psi\right)+\mathbf{k}\left(\frac{\partial\psi}{\partial t}+\frac{\partial\varphi}{\partial t}\cos\vartheta\right) \\ &= \mathbf{i}'\left(\frac{\partial\vartheta}{\partial t}\cos\varphi+\frac{\partial\psi}{\partial t}\sin\vartheta\sin\varphi\right) \\ &\quad+\mathbf{j}'\left(-\frac{\partial\vartheta}{\partial t}\sin\varphi+\frac{\partial\psi}{\partial t}\sin\vartheta\cos\varphi\right) \\ &\quad+\mathbf{k}'\left(\frac{\partial\varphi}{\partial t}+\frac{\partial\psi}{\partial t}\cos\vartheta\right) \end{aligned} \quad (181)$$

$$= \boldsymbol{\psi}_1 \frac{\partial \psi}{\partial t} + \boldsymbol{\vartheta}_1 \frac{\partial \vartheta}{\partial t} + \boldsymbol{\varphi}_1 \frac{\partial \varphi}{\partial t} = \left(\boldsymbol{\psi}_1 \frac{d\psi}{dt} + \boldsymbol{\vartheta}_1 \frac{d\vartheta}{dt} + \boldsymbol{\varphi}_1 \frac{d\varphi}{dt} \right)$$

$$+ \frac{d\psi}{dt} \vartheta(\boldsymbol{\psi}_1 \times \boldsymbol{\vartheta}_1) + \frac{d\psi}{dt} \varphi(\boldsymbol{\psi}_1 \times \boldsymbol{\varphi}_1) + \frac{d\vartheta}{dt} \varphi(\boldsymbol{\vartheta}_1 \times \boldsymbol{\varphi}_1)$$

$$= \frac{d}{dt} (\boldsymbol{\psi}_1 \psi + \boldsymbol{\vartheta}_1 \vartheta + \boldsymbol{\varphi}_1 \varphi).$$

The last member is a specially plain demand of the vector algebra, in order to reconcile the value of $(\dot{\boldsymbol{\gamma}})$ obtained by means of (XYZ) with the terms of equation (172) and its vector angles, and uphold the condition for invariant representation of $(\boldsymbol{\gamma})$ as the angular displacement proceeds. With this invariance put beyond critical doubt such vectors as $(\boldsymbol{\gamma})$ take their place under the procedure of equation (137), and we have detected here the earmarks of an invariant shift. A closer superficial agreement with that equation results from the coördination of derivatives calculated from equations (170, 171), because the axes (X′Y′Z′) remain orthogonal and *rotate*. With some watchful avoidance of confusion in the notation, the reasoning of section 80 can be duplicated, and the result confirmed without difficulty,

$$\dot{\boldsymbol{\gamma}} = \frac{d}{dt} (\mathbf{i}'\gamma_{(x')} + \mathbf{j}'\gamma_{(y')} + \mathbf{k}'\gamma_{(z')})$$

$$= \left[\mathbf{i}' \frac{d}{dt} (\gamma_{(x')}) + \mathbf{j}' \frac{d}{dt} (\gamma_{(y')}) + \mathbf{k}' \frac{d}{dt} (\gamma_{(z')}) \right] + (\dot{\boldsymbol{\gamma}} \times \boldsymbol{\gamma}), \quad (182)$$

where $(\dot{\boldsymbol{\gamma}})$ in the vector product must denote the shift rate for (X′Y′Z′), and the rest of that member shows the type of $(\dot{\mathbf{V}}_{(m)})$. We do not need now to transcribe the details of that development, with a less particular value for the shift rate.

98. Having made the beginning in section 93 with angular coördinate which may be placed in parallel with coördinate lengths, the above relation that introduces an angular velocity is liable to the same sort of double reading that was insisted upon in section 81, so that the change of reference-frame for angular velocity would also come to the front. Then using the third member of the last equation for illustration of a more general case, its first group can be said to present angular velocity relative to (X′Y′Z′), while the vector product added transfers correctly to (XYZ) as a standard. If this second branch of the idea is before us, a continuation of it in close likeness to the working out of consequences into equation (112) suggests itself naturally, in order to make a transfer between reference-frames that covers angular acceleration, as the previous equation provided for such a change in respect to linear accelerations. But that general provision will be omitted, with the intention of considering any special instance under its plan in the light of its own circumstances; and what attention is now to be given to angular acceleration will enter with the repetition of the one-step shift process, in which the original vector (**V**) is an angular velocity, and the derivative that appears in particular to replace the general derivative ($\dot{\mathbf{V}}$) of equation (137) is an angular acceleration, with the one standard frame retained, and no departures from invariant values finally tolerated.

That policy meets the requirements most frequently made in this field, and indeed the material that has grown to be classic and devoted to the relations of rotation-vectors and their derivatives to dynamical quantities, expressed especially by means of Euler's angles, marks its initial stage at the point that we have now reached. One feature of it, that we have once alluded to, is letting angle figure as an algebraic magnitude, but constructing a sequel where its two derivatives become vectors, effectively or with full recognition. It cannot be surprising, therefore, that those distinctions in respect to angular quantity, between its partial and its total time-derivative, nowhere need to appear in the classic equations; though we have been compelled to give them weight in the interest of correct work. Because both compensating elements in equations like (173) have their source in orientation, a view that excludes orientation needs neither; and the one magnitude derivative with respect to time that is retained may within certain limits raise no issue whether it is partial or total. There is however one place where comment has been the habit upon something of defect in the algebraic linkage, and where it is interesting to discover that the concept of vector angle does a little to make a better joint. We shall attempt to dispose of that minor matter in this pause between two steps of the more important progress.[1]

The comment in question hinges upon equations that the algebraic methods have always written equivalently to

$$\left.\begin{aligned} \mathbf{i}'\cdot\dot{\gamma} &= \frac{d\vartheta}{dt}\cos\varphi + \frac{d\psi}{dt}\sin\vartheta\sin\varphi; \\ \mathbf{j}'\cdot\dot{\gamma} &= -\frac{d\vartheta}{dt}\sin\varphi + \frac{d\psi}{dt}\sin\vartheta\cos\varphi; \\ \mathbf{k}'\cdot\dot{\gamma} &= \frac{d\varphi}{dt} + \frac{d\psi}{dt}\cos\vartheta; \end{aligned}\right\} \quad (183)$$

and where our sequences of thought have caused the substitution of time partials everywhere in the second members. If we pick out one equation for a sample, multiply by (dt) and write

$$(\mathbf{i}'\cdot\dot{\gamma})dt = d\vartheta\cos\varphi + d\psi\sin\vartheta\sin\varphi, \quad (184)$$

the usual and perfectly true remark about it and its companions is to this effect: The second members not being exact differentials under the ordinary test, because the equalities are not satisfied that would give for instance

$$\frac{\partial}{\partial\psi}(\cos\varphi) = \frac{\partial}{\partial\vartheta}(\sin\vartheta\sin\varphi), \quad (185)$$

[1] See Note 25.

there is some drawback upon using the first members. But if the vector plan retains the total derivatives in equations (183) and completes them, equation (184) becomes, as we have seen,

$$(\mathbf{i}' \cdot \dot{\gamma})dt = d\vartheta(\cos\varphi + \dot{\psi}\cos\vartheta\sin\varphi) + d\psi(\sin\vartheta\sin\dot{\varphi}) + d\varphi(-\dot{\vartheta}\sin\varphi + \dot{\psi}\sin\vartheta\cos\varphi), \quad (186)$$

in which the coefficients of $(d\vartheta, d\psi, d\varphi)$ do make the first member an exact differential by conforming to the standard rule, as direct test verifies. That particular drawback was removed by using vector angle in deriving the rotation-vector, and by aiming in our calculus deliberately to preserve the exact differentials that occurred naturally.

99. For the kind of inquiry that comes next in order, rotation-vectors in the standard frame are an assumed basis in the statement, being either given outright or brought within reach by such data related to Euler's angles as the foregoing sections have set forth. The undertaking looks toward expressing angular acceleration-vectors for the standard frame in terms of the same angles $(\boldsymbol{\psi}, \boldsymbol{\vartheta}, \boldsymbol{\varphi})$ and consequently in connection with some auxiliary frame like (X′Y′Z′). In its main outline this must stand as a parallel illustration of the method introduced before; but in order to vary from mere repetition, let there be one rotation-vector $(\boldsymbol{\omega})$ applying to a rigid body that is in pure rotation about the origin (O), and a second $(\dot{\boldsymbol{\gamma}})$ for the axes (X′Y′Z′), with whose aid $(\dot{\boldsymbol{\omega}})$ is to be determined through its projections upon them. We shall choose special assumptions, that will be found profitable because they anticipate one set of data met in real requirements of investigation. Let that definite line of the body, which is to have the angular coördinates $(\boldsymbol{\psi}, \boldsymbol{\vartheta})$ and thus specify those elements of the body's configuration, always coincide with (Z′); and to complete the assignment of relative configuration for body and axes, let $(\boldsymbol{\varphi})$ be permanently zero for the latter. Therefore (Y′) is contained permanently in the plane (Z′, Z), and (X′) in the normal to that plane. Distinguishing the angles applying to the axes as ($\boldsymbol{\psi}'$, $\boldsymbol{\vartheta}'$, $\boldsymbol{\varphi}'$) the conditions are

$$\boldsymbol{\psi}' = \boldsymbol{\psi}; \quad \boldsymbol{\vartheta}' = \boldsymbol{\vartheta}; \qquad \boldsymbol{\varphi}' = 0; \quad \boldsymbol{\varphi} \text{ (any value)}. \tag{187}$$

100. The rotation-vectors (ω) and ($\dot{\gamma}$) are now to be expressed, but that cannot be done by borrowing the forms from sections 95 and 97. For it is essential to the present circumstances that the sets of projections of each rotation-vector must give that quantity invariantly, as before it was exacted that the angle (γ) should be so expressed by equations (170, 171, 172). For every range in this use, equation (134) is to be made fundamental and characteristic. Going to one root of the matter in equations (111, 116), and holding to the leading thought of section 86, it becomes formally clear that no term like ($\dot{\gamma} \times \mathbf{V}$) of equation (137) can appear in forms adapted to the new independent start. And in reason it is convincing that projection at the moment is indifferent to past and future, and its results must be mathematically independent of a continuing process to which it is indifferent. All this fits perfectly our conception of each set (X′Y′Z′) as fixed, and ($\dot{\gamma}$) as a shift rate among the fixed sets. Bringing to equation (173) the modifying idea that ($\dot{\gamma}$) equal to zero must accompany the projection upon the individual set of axes for the epoch, we find first that the second groups in the coefficients of (**ijk**) drop away because they represent projections of a term like ($\dot{\gamma} \times \mathbf{V}$), and secondly that the difference between total and partial time-derivatives disappears in view of equations (178, 180). To be sure this detail is only a roundabout consequence of discarding at the one projection that which belongs only to a unified series of such projections as a whole; but it has bearing in dispelling lingering obscurities on the formal side of these matters. The point would not need to be labored so, were not misapprehension fostered by the misnomer *reference to moving axes* in speaking of them.

This is preface to writing the values

$$\omega = \psi_1 \frac{d\psi}{dt} + \vartheta_1 \frac{d\vartheta}{dt} + \phi_1 \frac{d\varphi}{dt}; \qquad \dot{\gamma} = \psi_1 \frac{d\psi}{dt} + \vartheta_1 \frac{d\vartheta}{dt}; \quad (188)$$

in order to proceed from them to the value of ($\dot{\omega}$) that is connected with the projections of (ω) on (X′Y′Z′). It seems worth noting that these may be corroborated by considering the particular configuration when (X′Y′Z′) fall in (XYZ), for which of course equality of projections must ensue. From equation (173) we see for that case and for the projections of (ω),

$$\dot{\gamma}_{(x)} = \mathbf{i}\,\frac{d\vartheta}{dt}; \qquad \dot{\gamma}_{(y)} = 0; \qquad \dot{\gamma}_{(z)} = \mathbf{k}\left(\frac{d\psi}{dt} + \frac{d\varphi}{dt}\right); \qquad (189)$$
$$\dot{\psi} = \dot{\vartheta} = \dot{\phi} = 0; \text{ and } \dot{\gamma} \equiv \omega.$$

It is true that the cancellations of terms arising from the type ($\dot{\gamma} \times \mathbf{V}$) now follow from ($\dot{\gamma}$) being zero, but they show consistency in the final outcome. The sum in ($\dot{\gamma}_{(z)}$) is contributed, part by turning of the plane (Y′Z) about (Z), and part by turning relatively to that plane about (Z′) coincident with (Z). Finally we can summarize in a brief rule the office of the two derivatives in connections like the present one: The partial time-derivative of the tensors enters where projection has preceded differentiation, and the total derivative where differentiation has preceded.

101. By projecting the rotation-vector (ω) upon (X′Y′Z′) we find

$$\left.\begin{aligned}
\omega_{(x')} &= \mathbf{i}'\,\frac{d\vartheta}{dt} = \vartheta_1 \frac{d\vartheta}{dt};\\
\omega_{(y')} &= \mathbf{j}'\left(\frac{d\psi}{dt}\sin\vartheta\right);\\
\omega_{(z')} &= \mathbf{k}'\left(\frac{d\varphi}{dt} + \frac{d\psi}{dt}\cos\vartheta\right) = \phi_1\left(\frac{d\varphi}{dt} + \frac{d\psi}{dt}\cos\vartheta\right);
\end{aligned}\right\} \quad (190)$$

the tensors being comprehensive or general values as explained in section 78, and therefore open to differentiation, whose execution yields

$$\left.\begin{aligned}
\frac{d}{dt}(\omega_{(x')}) &= \frac{d^2\vartheta}{dt^2};\\
\frac{d}{dt}(\omega_{(y')}) &= \frac{d^2\psi}{dt^2}\sin\vartheta + \frac{d\vartheta}{dt}\frac{d\psi}{dt}\cos\vartheta;\\
\frac{d}{dt}(\omega_{(z')}) &= \frac{d^2\varphi}{dt^2} + \frac{d^2\psi}{dt^2}\cos\vartheta - \frac{d\vartheta}{dt}\frac{d\psi}{dt}\sin\vartheta.
\end{aligned}\right\} \qquad (191)$$

The differentiation of equations (190) needs for its completion the terms introduced by changes of orientation in $(\mathbf{i'j'k'})$, which are

$$\left.\begin{aligned}
\dot{\boldsymbol{\vartheta}}_1\frac{d\vartheta}{dt} &= \frac{d\psi}{dt}(\boldsymbol{\psi}_1\times\boldsymbol{\vartheta}_1)\frac{d\vartheta}{dt} = (\boldsymbol{\psi}_1\times\boldsymbol{\vartheta}_1)\frac{d\psi}{dt}\frac{d\vartheta}{dt};\\
\dot{\mathbf{j}}'\frac{d\psi}{dt}\sin\vartheta &= \left(\frac{d\psi}{dt}(\boldsymbol{\psi}_1\times\mathbf{j}') + \frac{d\vartheta}{dt}(\boldsymbol{\vartheta}_1\times\mathbf{j}')\right)\frac{d\psi}{dt}\sin\vartheta\\
&= (\boldsymbol{\psi}_1\times\mathbf{j}')\left(\frac{d\psi}{dt}\right)^2\sin\vartheta + (\boldsymbol{\vartheta}_1\times\mathbf{j}')\frac{d\psi}{dt}\frac{d\vartheta}{dt}\sin\vartheta;\\
\dot{\boldsymbol{\varphi}}_1&\left(\frac{d\varphi}{dt} + \frac{d\psi}{dt}\cos\vartheta\right)\\
&= \left(\frac{d\psi}{dt}(\boldsymbol{\psi}_1\times\boldsymbol{\varphi}_1) + \frac{d\vartheta}{dt}(\boldsymbol{\vartheta}_1\times\boldsymbol{\varphi}_1)\right)\left(\frac{d\varphi}{dt} + \frac{d\psi}{dt}\cos\vartheta\right)\\
&= (\boldsymbol{\psi}_1\times\boldsymbol{\varphi}_1)\left(\frac{d\psi}{dt}\frac{d\varphi}{dt} + \left(\frac{d\psi}{dt}\right)^2\cos\vartheta\right)\\
&\qquad + (\boldsymbol{\vartheta}_1\times\boldsymbol{\varphi}_1)\left(\frac{d\vartheta}{dt}\frac{d\varphi}{dt} + \frac{d\vartheta}{dt}\frac{d\psi}{dt}\cos\vartheta\right).
\end{aligned}\right\} \qquad (192)$$

Next resolve the vector products into (X′, Y′, Z′) and assemble the terms for each one of the axes, which shows for the results when reduced by some cancellations

$$\left.\begin{aligned}
\dot{\omega}_{(x')} &= \boldsymbol{\vartheta}_1 \left(\frac{d^2\vartheta}{dt^2} + \frac{d\psi}{dt}\frac{d\varphi}{dt} \sin\vartheta \right); \\
\dot{\omega}_{(y')} &= \mathbf{j}' \left(\frac{d^2\psi}{dt^2} \sin\vartheta - \frac{d\vartheta}{dt}\frac{d\varphi}{dt} + \frac{d\psi}{dt}\frac{d\vartheta}{dt} \cos\vartheta \right); \\
\dot{\omega}_{(z')} &= \boldsymbol{\varphi}_1 \left(\frac{d^2\varphi}{dt^2} + \frac{d^2\psi}{dt^2} \cos\vartheta - \frac{d\psi}{dt}\frac{d\vartheta}{dt} \sin\vartheta \right).
\end{aligned}\right\} \quad (193)$$

In making the resolution the components of the vector products to be used are shown by

$$\begin{aligned}
&\boldsymbol{\psi}_1 \times \boldsymbol{\vartheta}_1 = \mathbf{j}' \cos\vartheta - \boldsymbol{\varphi}_1 \sin\vartheta; \qquad \boldsymbol{\psi}_1 \times \mathbf{j}' = -\boldsymbol{\vartheta}_1 \cos\vartheta; \\
&\boldsymbol{\vartheta}_1 \times \mathbf{j}' = \boldsymbol{\varphi}_1; \quad \boldsymbol{\psi}_1 \times \boldsymbol{\varphi}_1 = \boldsymbol{\vartheta}_1 \sin\vartheta; \quad \boldsymbol{\vartheta}_1 \times \boldsymbol{\varphi}_1 = -\mathbf{j}'.
\end{aligned} \quad (194)$$

Having obtained by these operations the projections of $(\dot{\omega})$ for the standard frame upon $(X'Y'Z')$, as corrected for the assumed shift of the axes, the total $(\dot{\omega})$ given by the vector sum of the second members is easily seen to be

$$\begin{aligned}
\dot{\omega} = \boldsymbol{\vartheta}_1 \frac{d^2\vartheta}{dt^2} + \boldsymbol{\psi}_1 \frac{d^2\psi}{dt^2} + \boldsymbol{\varphi}_1 \frac{d^2\varphi}{dt^2} + (\boldsymbol{\vartheta}_1 \times \boldsymbol{\varphi}_1) \frac{d\vartheta}{dt}\frac{d\varphi}{dt} \\
+ (\boldsymbol{\psi}_1 \times \boldsymbol{\vartheta}_1) \frac{d\psi}{dt}\frac{d\vartheta}{dt} + (\boldsymbol{\psi}_1 \times \boldsymbol{\varphi}_1) \frac{d\psi}{dt}\frac{d\varphi}{dt}.
\end{aligned} \quad (195)$$

And this last form of the value for the angular acceleration of the body is finally to be compared, on the one hand with the result of differentiating directly

$$\omega = \boldsymbol{\psi}_1 \frac{d\psi}{dt} + \boldsymbol{\vartheta}_1 \frac{d\vartheta}{dt} + \boldsymbol{\varphi}_1 \frac{d\varphi}{dt}, \quad (196)$$

and on the other hand, with the standard relation in equation (137). The first of these comparisons is no more than a matter of inspection, because the derivatives of the tensors appear immediately, and the known changes of orientation for $(\boldsymbol{\psi}_1, \boldsymbol{\vartheta}_1, \boldsymbol{\varphi}_1)$ are exactly accounted for in the vector products of equation (195). In order to carry through the other comparison we need for

($\dot{V}_{(m)}$) the derivatives of the tensors that are already recorded in equation (191), and whose vector sum can be thrown into the form, when the parts are duly oriented,

$$\dot{V}_{(m)} = \boldsymbol{\vartheta}_1 \frac{d^2\vartheta}{dt^2} + \boldsymbol{\psi}_1 \frac{d^2\psi}{dt^2} + \boldsymbol{\varphi}_1 \frac{d^2\varphi}{dt^2} + (\boldsymbol{\psi}_1 \times \boldsymbol{\vartheta}_1) \frac{d\psi}{dt}\frac{d\vartheta}{dt} \cdot \quad (197)$$

To this must be added

$$(\dot{\gamma} \times \omega) = \left(\boldsymbol{\psi}_1 \frac{d\psi}{dt} + \boldsymbol{\vartheta}_1 \frac{d\vartheta}{dt} \right) \times \left(\boldsymbol{\psi}_1 \frac{d\psi}{dt} + \boldsymbol{\vartheta}_1 \frac{d\vartheta}{dt} + \boldsymbol{\varphi}_1 \frac{d\varphi}{dt} \right), \quad (198)$$

whose expansion reduces to

$$(\dot{\gamma} \times \omega) = (\boldsymbol{\psi}_1 \times \boldsymbol{\varphi}_1) \frac{d\psi}{dt}\frac{d\varphi}{dt} + (\boldsymbol{\vartheta}_1 \times \boldsymbol{\varphi}_1) \frac{d\vartheta}{dt}\frac{d\varphi}{dt} ; \quad (199)$$

and confirms through the sum of equations (197, 199) the former value of ($\dot{\omega}$). Notice the difference in the segregation for the two groupings, by which the same term can be attributed at will to change of direction or of magnitude.

The components of (ω) in (XYZ) are obtainable in the forms

$$\left.\begin{aligned} \omega_{(x)} &= \mathbf{i} \left(\frac{d\vartheta}{dt} \cos\psi + \frac{d\varphi}{dt} \sin\vartheta \sin\psi \right); \\ \omega_{(y)} &= \mathbf{j} \left(\frac{d\vartheta}{dt} \sin\psi - \frac{d\varphi}{dt} \sin\vartheta \cos\psi \right); \\ \omega_{(z)} &= \mathbf{k} \left(\frac{d\psi}{dt} + \frac{d\varphi}{dt} \cos\vartheta \right); \end{aligned}\right\} \quad (200)$$

through which another plain road is opened to determine ($\dot{\omega}$); but we shall not go further here than to indicate it.

POLAR COÖRDINATES.

102. The system known as *polar coördinates* is a fitting sequel to what has just been done, because Euler's angles that we have denoted by ($\boldsymbol{\psi}, \boldsymbol{\vartheta}$) are universally employed to orient the radius-vector (**r**) whose pole is then taken at our origin (O). The angle

($\boldsymbol{\phi}$) is obviously superfluous when we are concerned with one line only and not with a body, even when ($\mathbf{r}$) moves in three dimensions; and when a limitation to the uniplanar conditions is imposed the pole is most often located in the plane of motion, and then of the three angles ($\boldsymbol{\psi}$) alone needs to be used. We shall guide the development toward the relations for three dimensions, and afterwards call attention to some briefer statements for the uniplanar case.

If we write the radius-vector ($\mathbf{r}$) as the product of its unit-vector ($\mathbf{r}_1$) and its tensor (r), according to one normal scheme of the vector algebra, the time-derivative ($\dot{\mathbf{r}}$) takes on the form

$$\dot{\mathbf{r}} = \mathbf{r}_1 \frac{dr}{dt} + \dot{\mathbf{r}}_1 r, \tag{201}$$

with unforced separation of the entire directional change from that which refers to the algebraic magnitude. By means of the results now at our disposal, the vector ($\dot{\boldsymbol{\gamma}}$) in application to the single line ($\mathbf{r}$) would lead straight to the expression for the velocity of (Q) at the extremity of ($\mathbf{r}$),

$$\mathbf{v} = \mathbf{r}_1 \frac{dr}{dt} + (\dot{\boldsymbol{\gamma}} \times \mathbf{r}) = \mathbf{r}_1 \frac{dr}{dt} + (\boldsymbol{\psi}_1 \times \mathbf{r}_1) r \frac{d\psi}{dt} + (\boldsymbol{\vartheta}_1 \times \mathbf{r}_1) r \frac{d\vartheta}{dt}. \tag{202}$$

From the second member, we infer at sight the truth of one usual statement about ($\mathbf{v}$): That it includes simultaneous motion on a sphere centered at the pole of ($\mathbf{r}$), and growth of ($\mathbf{r}$) in length. So long as we think strictly in the terms indicated, there is no rotation according to our use of that word; we deal with ($\dot{\boldsymbol{\gamma}}$) merely as the angular velocity of the one line. But when the third member of the last equation is drawn in, the set of axes (X′Y′Z′) as laid down in section 93 reappears, since the three parts of the velocity constitute always an orthogonal set, of which ($\mathbf{r}$) itself would be (Z′) in our adopted convention, coinciding with (Z) for zero values of ($\boldsymbol{\psi}$, $\boldsymbol{\vartheta}$). The completed consistent identification of axes and their true rotation-vector gives

$$\left.\begin{aligned} \mathbf{v}_{(z')} &= \mathbf{k}'\left(\frac{dr}{dt}\right); \qquad \mathbf{v}_{(y')} = \mathbf{j}'\left(\frac{d\psi}{dt}\, r \sin\vartheta\right); \\ \mathbf{v}_{(x')} &= \mathbf{i}'\left(\frac{d\vartheta}{dt}\, r\right); \\ \dot{\boldsymbol{\gamma}} &= \boldsymbol{\psi}_1\frac{d\psi}{dt} + \boldsymbol{\vartheta}_1\frac{d\vartheta}{dt}; \qquad \dot{\boldsymbol{\phi}} = 0 \text{ permanently.} \end{aligned}\right\} \quad (203)$$

It is self-evident that these three projections are an invariant equivalent for ($\mathbf{v}$), because they are in their source only the three parts of ($\dot{\mathbf{r}}$) in the standard frame. But we can also repeat the remark attached to equation (138), and enlarge it in the direction of presenting these polar coördinate relations for velocity in the light of a narrowly specialized instance within more elastic conditions.

Instead of binding (X′Y′Z′) to coincidence of (Z′) and ($\mathbf{r}$), let the axes rather move about the origin (O) as allowed by any general value of the rotation-vector ($\dot{\boldsymbol{\gamma}}$). The configuration of ($\mathbf{r}$) in the frame (X′Y′Z′) will be shown generally by

$$\mathbf{r} = \mathbf{i}'x' + \mathbf{j}'y' + \mathbf{k}'z'; \qquad (204)$$

and for those suppositions the general values of ($\dot{\mathbf{V}}$) and ($\dot{\mathbf{V}}_{(m)}$) in equation (137) will assume the form

$$\dot{\mathbf{r}} = \mathbf{r}_1\frac{dr}{dt} + (\boldsymbol{\psi}_1 \times \mathbf{r}_1)r\frac{d\psi}{dt} + (\boldsymbol{\vartheta}_1 \times \mathbf{r}_1)r\frac{d\vartheta}{dt}$$

$$= \mathbf{i}'\frac{dx'}{dt} + \mathbf{j}'\frac{dy'}{dt} + \mathbf{k}'\frac{dz'}{dt} + (\dot{\boldsymbol{\gamma}} \times \mathbf{r}). \qquad (205)$$

The effect of that particular choice for the rotation-vector in equation (203) is then put clearly in evidence: the velocity of (Q) at the extremity of ($\mathbf{r}$), but reckoned relatively to the frame (X′Y′Z′), is thrown exclusively upon the axis (Z′), while (x′, y′) remain permanently at zero, and the term ($\dot{\boldsymbol{\gamma}} \times \mathbf{r}$) is left to bring in all of both components that (**v**) shows parallel to (X′) and to (Y′). Or in the alternative reading, the correction for shift of orientation being perpendicular to (**r**), it is segregated completely from the only change in tensor magnitude that is allowed to become realized in (X′Y′Z′).

103. The natural order proceeds next to take up, with polar variables as instruments, the task of expressing the polar components of the acceleration with which (Q) moves relatively to the standard frame, and which can be determined otherwise, as we know, by projecting the resultant ($\dot{\mathbf{v}}$) upon the directions of (X′Y′Z′) at the epoch. However these projections may be written originally, the translation into functions of ($\mathbf{r}$, $\boldsymbol{\psi}$, $\boldsymbol{\vartheta}$) is a matter of algebra only. Leaving that method aside, the details will be worked out in two ways, both moving with reasonable directness toward the end in view, and each having its own interest through the vector algebra of it. Let us carry out first the application of equation (137). It gives

$$\left.\begin{aligned} \dot{\mathbf{V}}_{(m)} &= \mathbf{i}'\left(r\frac{d^2\vartheta}{dt^2} + \frac{d\vartheta}{dt}\frac{dr}{dt}\right) \\ &\quad + \mathbf{j}'\left(r\frac{d^2\psi}{dt^2}\sin\vartheta + \frac{dr}{dt}\frac{d\psi}{dt}\sin\vartheta + \frac{d\psi}{dt}\frac{d\vartheta}{dt}r\cos\vartheta\right) \\ &\quad + \mathbf{k}'\left(\frac{d^2r}{dt^2}\right); \\ (\dot{\boldsymbol{\gamma}} \times \mathbf{v}) &= \left[\boldsymbol{\psi}_1\frac{d\psi}{dt} + \boldsymbol{\vartheta}_1\frac{d\vartheta}{dt}\right] \\ &\quad \times \left[\mathbf{r}_1\frac{dr}{dt} + (\boldsymbol{\psi}_1 \times \mathbf{r}_1)r\frac{d\psi}{dt} + (\boldsymbol{\vartheta}_1 \times \mathbf{r}_1)r\frac{d\vartheta}{dt}\right]. \end{aligned}\right\} \quad (206)$$

As a help in expanding the second equation these relations enter:

$$\left.\begin{aligned} (\boldsymbol{\psi}_1 \times \mathbf{r}_1) &= \boldsymbol{\vartheta}_1 \sin\vartheta; \\ \boldsymbol{\psi}_1 \times (\boldsymbol{\psi}_1 \times \mathbf{r}_1) &= -\,\mathbf{i}'\sin\vartheta\cos\vartheta - \mathbf{r}_1\sin^2\vartheta; \\ \boldsymbol{\psi}_1 \times (\boldsymbol{\vartheta}_1 \times \mathbf{r}_1) &= \boldsymbol{\vartheta}_1\cos\vartheta; \qquad (\boldsymbol{\vartheta}_1 \times \mathbf{r}_1) = \mathbf{i}'; \\ \boldsymbol{\vartheta}_1 \times (\boldsymbol{\psi}_1 \times \mathbf{r}_1) &= 0; \qquad \boldsymbol{\vartheta}_1 \times (\boldsymbol{\vartheta}_1 \times \mathbf{r}_1) = -\,\mathbf{r}_1. \end{aligned}\right\} \quad (207)$$

Summing the items in their proper orientation, the polar components of ($\dot{\mathbf{v}}$) are found to be

$$\left.\begin{aligned}\dot{\mathbf{v}}_{(x')} &= \mathbf{i}'\left(\mathrm{r}\frac{d^2\vartheta}{dt^2} + 2\frac{dr}{dt}\frac{d\vartheta}{dt} - \mathrm{r}\left(\frac{d\psi}{dt}\right)^2 \sin\vartheta\cos\vartheta\right);\\ \dot{\mathbf{v}}_{(y')} &= \mathbf{j}'\left(\mathrm{r}\frac{d^2\psi}{dt^2}\sin\vartheta + 2\frac{d\psi}{dt}\frac{dr}{dt}\sin\vartheta + 2\mathrm{r}\frac{d\vartheta}{dt}\frac{d\psi}{dt}\cos\vartheta\right);\\ \dot{\mathbf{v}}_{(z')} &= \mathbf{k}'\left(\frac{d^2r}{dt^2} - \mathrm{r}\left(\frac{d\vartheta}{dt}\right)^2 - \mathrm{r}\left(\frac{d\psi}{dt}\right)^2\sin^2\vartheta\right).\end{aligned}\right\} \quad (208)$$

The second development picks up its thread at equation (201), and differentiates that again as it stands; so the first stage shows immediately

$$\ddot{\mathbf{r}} \equiv \dot{\mathbf{v}} = \mathbf{r}_1\frac{d^2r}{dt^2} + 2\dot{\mathbf{r}}_1\frac{dr}{dt} + \ddot{\mathbf{r}}_1\mathrm{r}; \quad (209)$$

and carrying out some of the indicated operations yields

$$\left.\begin{aligned}\dot{\mathbf{r}}_1 &= (\dot{\boldsymbol{\gamma}} \times \mathbf{r}_1);\\ \ddot{\mathbf{r}}_1 &= (\ddot{\boldsymbol{\gamma}} \times \mathbf{r}_1) + (\dot{\boldsymbol{\gamma}} \times \dot{\mathbf{r}}_1) = (\ddot{\boldsymbol{\gamma}} \times \mathbf{r}_1) + (\dot{\boldsymbol{\gamma}} \times (\dot{\boldsymbol{\gamma}} \times \mathbf{r}_1));\\ \ddot{\boldsymbol{\gamma}} &= \dot{\boldsymbol{\psi}}_1\frac{d\psi}{dt} + \boldsymbol{\psi}_1\frac{d^2\psi}{dt^2} + \dot{\boldsymbol{\vartheta}}_1\frac{d\vartheta}{dt} + \boldsymbol{\vartheta}_1\frac{d^2\vartheta}{dt^2};\\ &\text{with } \boldsymbol{\psi}_1 \text{ constant, } \dot{\boldsymbol{\vartheta}}_1 = \frac{d\psi}{dt}(\boldsymbol{\psi}_1 \times \boldsymbol{\vartheta}_1);\\ (\ddot{\boldsymbol{\gamma}} \times \mathbf{r}_1) &= (\boldsymbol{\psi}_1 \times \mathbf{r}_1)\frac{d^2\psi}{dt^2} + \left((\boldsymbol{\psi}_1 \times \boldsymbol{\vartheta}_1)\frac{d\psi}{dt}\frac{d\vartheta}{dt}\right) \times \mathbf{r}_1\\ &\quad + (\boldsymbol{\vartheta}_1 \times \mathbf{r}_1)\frac{d^2\vartheta}{dt^2};\\ \dot{\boldsymbol{\gamma}} \times (\dot{\boldsymbol{\gamma}} \times \mathbf{r}_1) &= [\boldsymbol{\psi}_1 \times (\boldsymbol{\psi}_1 \times \mathbf{r}_1)]\left(\frac{d\psi}{dt}\right)^2\\ &\quad + [\boldsymbol{\psi}_1 \times (\boldsymbol{\vartheta}_1 \times \mathbf{r}_1)]\frac{d\psi}{dt}\frac{d\vartheta}{dt} + [\boldsymbol{\vartheta}_1 \times (\boldsymbol{\psi}_1 \times \mathbf{r}_1)]\frac{d\vartheta}{dt}\frac{d\psi}{dt}\\ &\quad + [\boldsymbol{\vartheta}_1 \times (\boldsymbol{\vartheta}_1 \times \mathbf{r}_1)]\left(\frac{d\vartheta}{dt}\right)^2.\end{aligned}\right\} \quad (210)$$

Substituting these values in equation (209), it is recognizable readily with the aid of equations (207) that the results of the two methods are in perfect agreement.

104. The adjustment of the foregoing analysis to the simplified conditions of uniplanar motion, where the pole for ($\mathbf{r}$) is taken in the plane of the motion, will make (ϑ) constantly a right angle, so that ($\mathbf{r}$) revolves in the equatorial plane of the sphere whose polar axis is (Z). In adaptation to that case the velocity components are

$$\mathbf{v}_{(z')} = \mathbf{v}_{(r)} = \mathbf{r}_1 \frac{dr}{dt}; \qquad \mathbf{v}_{(y')} = j'\left(r\frac{d\psi}{dt}\right); \qquad \mathbf{v}_{(x')} = 0; \quad (211)$$

and the acceleration components become

$$\left.\begin{aligned} \dot{\mathbf{v}}_{(z')} &= \dot{\mathbf{v}}_{(r)} = \mathbf{r}_1\left(\frac{d^2r}{dt^2} - r\left(\frac{d\psi}{dt}\right)^2\right); \\ \dot{\mathbf{v}}_{(y')} &= j'\left(r\frac{d^2\psi}{dt^2} + 2\frac{d\psi}{dt}\frac{dr}{dt}\right); \\ \dot{\mathbf{v}}_{(x')} &= 0. \end{aligned}\right\} \quad (212)$$

Even on this simpler level, and after removing those complications which belong to the freedom in three dimensions, the same feature remains prominent through all the results; in one sense the idea of superposition fails. For though the resultant velocity contains neither more nor less than the parts due to the radial motion by itself and the revolution by itself, we cannot build up in that fashion the acceleration ($\dot{\mathbf{v}}$) of equations (208), nor yet of equations (212). In the latter, the second term in the coefficient of (j′) does not belong to the radial motion, nor to the circular motion, but it appears only when these two types coexist. And under the broader conditions, the coexistence in pairs of the three component velocities asserts itself through the terms in the acceleration:

$$i'\left(2\frac{dr}{dt}\frac{d\vartheta}{dt}\right); \qquad j'\left(2\frac{dr}{dt}\frac{d\psi}{dt}\right); \qquad j'\left(2r\frac{d\vartheta}{dt}\frac{d\psi}{dt}\right). \quad (213)$$

In view of their obtrusive symmetry, it is somewhat surprising that the force depending on the third of the group should have invited and fixed nearly exclusive attention: it is the famous *compound centrifugal force* with which the name of Coriolis has been associated.[1]

Approaching along the line now laid down to follow, these terms can be traced intelligently to a common origin in the nature of the coördinate system that is being employed; their appearance is connected essentially with the changes of direction peculiar to the descriptive vectors that are used. On that side, the parts of the force that match such accelerations may be declared mathematical, though it must be granted that they can become sound physics too, whenever those descriptive vectors are closely fitted to the physical action. In a centrifugal pump, a force that goes with the coefficient of (j′) above does work and strains the structural parts. But the same term shows in the algebra, when constant velocity is referred to a pole lying outside the straight line path, although no net force at all can then be active. It is also a significant fact that the factor (2) in each case makes its appearance because two terms coalesce, whose function is different in respect to the vector quantities that they affect. It is half-and-half change of magnitude in one vector and change of direction in a second distinct vector, as our process of derivation demonstrates. So the force of Coriolis cannot give a definitive account of gyroscopic phenomena on the basis of an incident in the algebra; first, it must be exhibited to correspond with traceable dynamical action. The same lesson is enforced here as by the matters broached in sections 35 and 57, of which the latter is peculiarly pertinent in that it brings forward the idea that angular acceleration, and therefore the coexistence of rotations about (ψ_1) and (ϑ_1) that is characteristic of the compound centrifugal force, may come about

[1] See Note 26.

in the absence of all force-moment, as a symptom that control is absent, not that it is present and is producing these effects.

105. The general values of equation (208) cover as a special case, it is plain, the condition that (**r**) shall be constant in length which goes with a pure rotation about (O). Consequently if we make that assumption here, the special value of ($\dot{\mathbf{v}}$) that is obtained must be reconcilable with the determination made in sections 54 and 101. Only the latter, in its turn, must be specialized for a point situated in its axis of ($\dot{\boldsymbol{\phi}}_1$), which is now also that of ($\mathbf{r}_1$). The notation in the two sections is consistent with the same supposition about the rotation-vector ($\dot{\boldsymbol{\gamma}}$) of (X′Y′Z′); and the axis (Z′) is common to both inquiries. But it will be observed that ($\boldsymbol{\vartheta}_1$) of section (101) is identified with (i′), and ($\boldsymbol{\vartheta}_1$) of equations (203) is paired with (j′); and hence a comparison of results must adopt in correspondence

(i′);	(j′);	(**k**′);	[Equations (193)]
(j′);	(− i′);	(**k**′);	[Equations (208)]

in order to preserve the right-handed cycle.

If (**r**) is constant in length the terms remaining in equations (208) are

$$\left.\begin{aligned} \dot{\mathbf{v}}_{(x')} &= \mathrm{i}'\left(\mathrm{r}\frac{d^2\vartheta}{dt^2} - \mathrm{r}\left(\frac{d\psi}{dt}\right)^2 \sin\vartheta\cos\vartheta\right); \\ \dot{\mathbf{v}}_{(y')} &= \mathrm{j}'\left(\mathrm{r}\frac{d^2\psi}{dt^2}\sin\vartheta + 2\mathrm{r}\frac{d\psi}{dt}\frac{d\vartheta}{dt}\cos\vartheta\right); \\ \dot{\mathbf{v}}_{(z')} &= \mathbf{k}'\left(-\mathrm{r}\left(\frac{d\vartheta}{dt}\right)^2 - \mathrm{r}\left(\frac{d\psi}{dt}\right)^2\sin^2\vartheta\right). \end{aligned}\right\} \tag{214}$$

And the vector sum of these must agree with equation (72) after the latter has been adapted to the point

$$z' = r; \qquad x' = y' = 0. \tag{215}$$

We have for use with equations (72, 188, 193)

$$\left.\begin{aligned}
&(\dot{\boldsymbol{\omega}} \times \mathbf{r}) = \mathbf{i}'(\dot{\omega}_{(\mathbf{y}')}\mathrm{r}) - \mathbf{j}'(\dot{\omega}_{(\mathbf{x}')}\mathrm{r});\\
&\boldsymbol{\omega}(\boldsymbol{\omega}\cdot\mathbf{r}) - \mathbf{r}(\omega^2)\\
&\quad = \left[\boldsymbol{\psi}_1 \frac{\mathrm{d}\psi}{\mathrm{dt}} + \boldsymbol{\vartheta}_1 \frac{\mathrm{d}\vartheta}{\mathrm{dt}} + \boldsymbol{\phi}_1 \frac{\mathrm{d}\varphi}{\mathrm{dt}}\right]\left(\mathrm{r}\left(\frac{\mathrm{d}\psi}{\mathrm{dt}} \cos\vartheta + \frac{\mathrm{d}\varphi}{\mathrm{dt}}\right)\right)\\
&\quad - \mathbf{r}\left(\left(\frac{\mathrm{d}\psi}{\mathrm{dt}}\right)^2 + \left(\frac{\mathrm{d}\vartheta}{\mathrm{dt}}\right)^2 + \left(\frac{\mathrm{d}\varphi}{\mathrm{dt}}\right)^2 + 2\frac{\mathrm{d}\psi}{\mathrm{dt}}\frac{\mathrm{d}\varphi}{\mathrm{dt}}\cos\vartheta\right).
\end{aligned}\right\} \quad (216)$$

When the multiplications are carried out and the items duly oriented by the plan explicitly recognized for equations (214), the values are found in agreement at all points.

The special circumstances to which equations (214) conform, make them express the acceleration of a point in the symmetry-axis of a top or gyroscope when it is spinning about that axis while the latter is executing any motions that change ($\boldsymbol{\vartheta}$) and ($\boldsymbol{\psi}$). Beside the utility of this value in application to the problem of the top, and the consolidation that the conclusions attain through the comparison, it is particularly instructive to follow carefully and in detail the appearance of terms in the acceleration, and their various disappearances by cancellation. Then one learns to cross-examine the mathematics and to discount sensibly its evidence or suggestion as to just what dynamical processes are in operation.

106. The fact that the resolution into polar component shapes itself in accommodation to each individual radius-vector prevents the introduction of any usefully general integrations to include extended masses. As a substitute recourse is had, where the radius-vector enters naturally, to plans like that worked out for the rotation of a rigid body, which has contrived to extract the common elements ($\boldsymbol{\omega}$) and ($\dot{\boldsymbol{\omega}}$) for use with all radius-vectors, and the moments of inertia as factors that cover the whole mass. The polar components that have been deduced are then limited practically to one mass-element or to the particle at the center of mass of the body. For the latter case, there is no difficulty in writing down for the six fundamental quantities the parts of their standard frame values that match the orthogonal polar projections. These are:

$$\left.\begin{aligned}
\mathbf{Q} &= m\left[(\boldsymbol{\vartheta}_1 \times \mathbf{r}_1)r\frac{d\vartheta}{dt} + (\boldsymbol{\psi}_1 \times \mathbf{r}_1)r\frac{d\psi}{dt} + \mathbf{r}_1\frac{dr}{dt}\right] \\
&\qquad = \mathbf{Q}_{(x')} + \mathbf{Q}_{(y')} + \mathbf{Q}_{(z')}; \\
E &= \tfrac{1}{2}m[v^2_{(x')} + v^2_{(y')} + v^2_{(z')}] \\
&\qquad = E_{(x')} + E_{(y')} + E_{(z')}; \\
\mathbf{H} &= m[-\mathbf{i}'(rv_{(y')}) + \mathbf{j}'(rv_{(x')})] \\
&\qquad = \mathbf{H}_{(x')} + \mathbf{H}_{(y')}; \quad [\mathbf{H}_{(z')} = 0]; \\
\mathbf{R} &= m[\mathbf{i}'\dot{v}_{(x')} + \mathbf{j}'\dot{v}_{(y')} + \mathbf{k}'\dot{v}_{(z')}] \\
&\qquad = \mathbf{R}_{(x')} + \mathbf{R}_{(y')} + \mathbf{R}_{(z')}; \\
P &= R_{(x')}v_{(x')} + R_{(y')}v_{(y')} + R_{(z')}v_{(z')}; \\
\mathbf{M} &= -\mathbf{i}'(R_{(y')}) + \mathbf{j}'(rR_{(x')}) = \mathbf{M}_{(x')} + \mathbf{M}_{(y')}; \\
&\qquad [\mathbf{M}_{(z')} = 0].
\end{aligned}\right\} \quad (217)$$

As an addendum to the separation of power or activity (P) into its parts it is worth noting that the total force corresponding to the heading $(\dot{\boldsymbol{\gamma}} \times \mathbf{v})$ of equation (206) can finally contribute nothing to the work done. It must of necessity be perpendicular to $(\mathbf{v})$ and therefore ineffective in the product $(\mathbf{R}\cdot\mathbf{v})$. Amounts of work per second may be yielded in the parts of (P) by the inclusion of these *directional forces*, but they must be self-compensating and give zero of work in the aggregate. Their behavior in both respects toward power is similar to that of normal force that is confined to changing direction in resultant momentum. Under $(\dot{\mathbf{V}}_{(m)})$, other elements of force may be entered that also give change of direction to $(m\mathbf{v})$; this function it may share with $(\dot{\boldsymbol{\gamma}} \times \mathbf{v})$. But $(\dot{\mathbf{V}}_{(m)})$ has monopoly, as was pointed out earlier, of bringing about all changes of magnitude in (**v**), and hence in (m**v**). It is plain common sense to confirm these conclusions by the observation that what happens to coördinates merely—to the descriptive vectors as we have called them—cannot affect the physical data that they are devised to describe.

Hansen's Ideal Coördinates.

107. By the trend of the standard illustrations, it cannot fail to have grown conspicuous already, how varied the available combinations must be and how many kinds of adjustment to special purposes are rendered possible, when once such resources and expedients have been brought under fair control, and a definite formulation of the ends sought has been arrived at. The next instance in order, the *ideal coördinates* so named by Hansen who proposed them, is adapted to strengthen that perception.[1] The invention of the plan seems to have been consciously directed by a purpose, and it finds a place here because it has made its standing good for certain fields of application. As would be natural to surmise, the proposals that have won acceptance have been gleaned by the sifting of actual and continued trial among the larger number submitted for general approval. Ideal coördinates are made to follow upon the polar system here because the radius-vector still remains a prominent element in their specifications; and on this account, they too have no immediate range beyond tracing the motion of one particle or mass-element. It will be recognized that they pursue, like the other coördinate systems that have been discussed, the object of stating standard frame values, but in more elastic partition of the totals than (XYZ) itself can furnish.

The chief concern of ideal coördinates is with velocity, and its main course may be called a response to the question, in what direction can the restrictions upon the frame (X'Y'Z'), that the

[1] See Note 27.

polar system has been seen to impose, be loosened without impairing the invariance of ($\mathbf{v}$) that the polar components retain. That point being secured, the other consequences entailed are left in whatever form they may happen to appear. In this way it becomes part of the inquiry to ascertain how the expression of acceleration is affected by the assumed conditions. The frames (X′Y′Z′) and (XYZ) continue with a common origin (O).

108. If we add to the suppositions of section 102 a rotation of (X′Y′Z′) about (Z′) that can be of any assigned magnitude, equation (202) will be written, when as before we identify ($\boldsymbol{\phi}_1$, $\mathbf{k}'$, and $\mathbf{r}_1$),

$$\mathbf{v} = \mathbf{r}_1 \frac{dr}{dt} + (\boldsymbol{\psi}_1 \times \mathbf{r}_1) r \frac{d\psi}{dt} + (\boldsymbol{\vartheta}_1 \times \mathbf{r}_1) r \frac{d\vartheta}{dt} + (\boldsymbol{\phi}_1 \times \mathbf{r}_1) r \frac{d\varphi}{dt}\,; \tag{218}$$

but the difference introduced is only formal since ($\boldsymbol{\phi}_1$, $\mathbf{r}_1$) are identical unit-vectors, and in this frame (X′Y′Z′) it is still the coördinate ($\mathbf{r}$) or ($\mathbf{z}'$) alone that can differ from zero, while the same corrections make the previous invariant representation of ($\mathbf{v}$) persist. This puts before us the nucleus of Hansen's idea, as vector algebra allows us to condense it. Now it will not be overlooked that ($\mathbf{v}_{(x')}$, $\mathbf{v}_{(y')}$, $\mathbf{v}_{(z')}$) as determined by equation (203) are the components of ($\mathbf{v}$) in that frame of permanent configuration in (XYZ) for which, with ($\boldsymbol{\phi}$) equal to zero, the frame (X′Y′Z′) is the indicator at the epoch. But it follows from the form of equation (218) that a whole group of fixed frames which at the epoch have (Z′) in common and are distributed through all azimuths round that axis for the range (0, 2π) in (φ), satisfy first the relation for the vector sum

$$\mathbf{v}_{(x')} + \mathbf{v}_{(y')} = (\boldsymbol{\psi}_1 \times \mathbf{r}_1) r \frac{d\psi}{dt} + (\boldsymbol{\vartheta}_1 \times \mathbf{r}_1) r \frac{d\vartheta}{dt}\,, \tag{219}$$

and accordingly for the invariance of

$$\mathbf{v} = \mathbf{v}_{(x')} + \mathbf{v}_{(y')} + \mathbf{v}_{(z')}. \tag{220}$$

Whatever the direction therefore, in which the extremity of ($\mathbf{r}$) is instantaneously moving parallel to the (X′Y′) plane, it is possible to select at that epoch among the group mentioned above one frame for which ($\mathbf{v}_{(x')}$) is zero, and another for which ($\mathbf{v}_{(y')}$) is zero; and whichever alternative is chosen of these two it is further open to attempt determining the rate of the rotation about (Z′) so that this one component remains permanently zero. We shall return presently and develop consequences of those possibilities, after pausing to insist a little upon equation (220) which has not yet been particularized in that sense.

109. In order to come nearer to the form of statement that Hansen was compelled to employ, go back to section 89, where equations (150, 151) express the invariance of ($\mathbf{r}$) in frames having a common origin. Let us pass on to consider equations (154), noticing how the added invariance of ($\mathbf{v}$) necessitates the vanishing of the last group of terms in each of them, for which one condition extracted from equation (162) is seen to be that ($\dot{\gamma}$) though differing from zero is colinear with ($\mathbf{r}$). For our benefit just now, this signifies that if two frames give equivalent sets of components for the same resultant velocity, the equivalence will not be disturbed by allowing one of them to be subject to a shift, provided that the axis of it lies in the radius-vector at the epoch. Then, as Hansen puts it, equations (151, 154) will exhibit the same type in their forms, with velocities replacing everywhere the corresponding coördinates, and the ideal for (x′y′z′) has been reached. As we have approached it there are two stages: the shift of (X′Y′Z′) in the angular coördinates (ψ, ϑ) is not without influence upon the relations, but it has been compensated in equations (203), and adding then a supplementary shift about ($\mathbf{r}_1$) that is also (φ_1) leaves this compensation untouched.

The zero value of (φ) having been standardized for equation (203) with (X′) in the plane (Z′, Z), for the more general value of (φ) that is now contemplated we should write

$$\begin{aligned} \mathbf{v}_{(x')} &= \mathbf{i}'\left(r\frac{d\vartheta}{dt}\cos\varphi + r\frac{d\psi}{dt}\sin\vartheta\sin\varphi\right); \\ \mathbf{v}_{(y')} &= \mathbf{j}'\left(-r\frac{d\vartheta}{dt}\sin\varphi + r\frac{d\psi}{dt}\sin\vartheta\cos\varphi\right). \end{aligned} \tag{221}$$

And if we settle upon making ($\mathbf{v}_{(y')}$) zero, the proper value of ($\boldsymbol{\varphi}$) at the epoch is determined by

$$\operatorname{tg}\varphi' = \frac{r\dfrac{d\psi}{dt}\sin\vartheta}{r\dfrac{d\vartheta}{dt}}. \tag{222}$$

Let us retain ($\dot{\boldsymbol{\gamma}}$) for the rotation-vector of (X′Y′Z′), and distinguish by ($\boldsymbol{\omega}$) the angular velocity of ($\mathbf{r}$), so that in the subsequent details

$$\dot{\boldsymbol{\gamma}} = \boldsymbol{\psi}_1\frac{d\psi}{dt} + \boldsymbol{\vartheta}_1\frac{d\vartheta}{dt} + \boldsymbol{\varphi}_1\frac{d\varphi}{dt}; \qquad \boldsymbol{\omega} = \boldsymbol{\psi}_1\frac{d\psi}{dt} + \boldsymbol{\vartheta}_1\frac{d\vartheta}{dt}. \tag{223}$$

Then under the condition of adjustment shown by equation (222) we have

$$\begin{aligned} \mathbf{v}_{(x')} &= (\boldsymbol{\omega}\times\mathbf{r}) = \mathbf{i}'\left[r^2\left(\frac{d\vartheta}{dt}\right)^2 + r^2\left(\frac{d\psi}{dt}\right)^2\sin^2\vartheta\right]^{\frac{1}{2}}; \\ \mathbf{v}_{(y')} &= 0; \qquad \mathbf{v}_{(z')} = \mathbf{r}_1\frac{dr}{dt}. \end{aligned} \tag{224}$$

110. The execution of this manoeuvre reduces the statement, so far as velocities are concerned, to one of motion in an instantaneously oriented plane (Z′X′), with a resolution of ($\mathbf{v}$) for the standard frame along the radius-vector and the perpendicular to it in that plane. The values of the components conform perfectly in type to those of the similar projections in the permanent plane of uniplanar conditions; and the prospect is opened for success in determining such a rate of rotation about ($\mathbf{r}$) as will perpetuate the instantaneous relations in exactly this form when they have been established at some one epoch; this involves keeping the values of ($\mathbf{v}_{(y')}$) continuously at zero, though it is always reckoned in the normal to the shifting plane (Z′X′). The examination of the arrangement requisite to that end is connected with the question about components of the acceleration ($\dot{\mathbf{v}}$), and we shall make our beginning there.

Recorded in equations (208) are the projections of ($\dot{\mathbf{v}}$) for (XYZ) upon the (X′Y′Z′) axes as located by ($\boldsymbol{\phi} = 0$); and from them can be calculated the equivalent set of projections upon the axes (X′Y′Z′) located by the general value of ($\boldsymbol{\phi}$), precisely as equation (221) does this for velocity. Those projections can finally be particularized for the angle ($\boldsymbol{\phi}'$) assigned by equation (222) to satisfy its announced condition. Distinguish the last named components of ($\dot{\mathbf{v}}$) temporarily as ($\dot{\mathbf{v}}_{(x'')}$, $\dot{\mathbf{v}}_{(y'')}$, $\dot{\mathbf{v}}_{(z'')}$); they are given by

$$\left.\begin{aligned} \dot{\mathbf{v}}_{(x'')} &= \mathbf{i}''(\dot{v}_{(x')} \cos \varphi' + \dot{v}_{(y')} \sin \varphi'); \\ \dot{\mathbf{v}}_{(y'')} &= \mathbf{j}''(- \dot{v}_{(x')} \sin \varphi' + \dot{v}_{(y')} \cos \varphi'); \\ \dot{\mathbf{v}}_{(z'')} &= \mathbf{k}''(\dot{v}_{(z')}); \quad \text{with } \mathbf{k}'' = \mathbf{k}' = \mathbf{r}_1, \\ &\text{the new unit-vectors being } (\mathbf{i}''\mathbf{j}''\mathbf{k}''). \end{aligned}\right\} \tag{225}$$

In the text of section 103, the components of ($\dot{\mathbf{v}}$) happen to find expression through polar variables, but that is plainly only an incident of the sequence in which they were developed; they might just as well have been derived from

$$\left.\begin{aligned} \dot{\mathbf{v}}_{(x'')} &= \mathbf{i}''\left(\mathbf{i}''\cdot\mathbf{i}\frac{d^2x}{dt^2} + \mathbf{i}''\cdot\mathbf{j}\frac{d^2y}{dt^2} + \mathbf{i}''\cdot\mathbf{k}\frac{d^2z}{dt^2}\right); \\ \dot{\mathbf{v}}_{(y'')} &= \mathbf{j}''\left(\mathbf{j}''\cdot\mathbf{i}\frac{d^2x}{dt^2} + \mathbf{j}''\cdot\mathbf{j}\frac{d^2y}{dt^2} + \mathbf{j}''\cdot\mathbf{k}\frac{d^2z}{dt^2}\right); \\ \dot{\mathbf{v}}_{(z'')} &= \mathbf{k}''\left(\mathbf{k}''\cdot\mathbf{i}\frac{dx^2}{dt^2} + \mathbf{k}''\cdot\mathbf{j}\frac{d^2y}{dt^2} + \mathbf{k}''\cdot\mathbf{k}\frac{d^2z}{dt^2}\right); \end{aligned}\right\} \tag{226}$$

or in some other equivalent fashion, the choice depending upon how the data are presented. It is another consequence of this idea, that the original shift of (X′Y′Z′) in (ψ, ϑ) belonging to polar components is unessential; in effect it drops out of consideration through the allowances for its presence when equations (208) were made correct, as we saw also in speaking about equations (203). The vital element in these ideal coördinates is the accompanying rotation about (**r**) which has been relied on at critical points to secure at once invariance and simplification in the relations for the velocities, and whose consistent introduction into those for acceleration we are now prepared to finish. In order to accentuate the real dissociation from the polar scheme, let us think definitely in the terms suggested for equations (226), of two coincident frames in the configuration with (XYZ) designated by (ψ, ϑ, ϕ'), of which one is fixed, while the other is departing from coincidence by rotation about the (**r**) of the epoch. We will temporarily call the rate of this departure (**u**) in substitution for the time rate (ϕ_1 dφ/dt).

111. Then the specialization of equation (137) to these circumstances gives, if we particularize the velocities also as ($\mathbf{v}_{(x'')}$, $\mathbf{v}_{(y'')}$, $\mathbf{v}_{(z'')}$), and remember

$$\mathbf{u}_{(z'')} = \mathbf{u}; \qquad \mathbf{u}_{(x'')} = \mathbf{u}_{(y'')} = 0; \qquad \mathbf{v}_{(y'')} = 0; \qquad (227)$$

$$\left.\begin{aligned} \dot{\mathbf{v}}_{(x'')} &= \mathbf{i}'' \frac{d}{dt}(\mathrm{v}_{(x'')}); \\ \dot{\mathbf{v}}_{(y'')} &= \mathbf{j}'' \left[\frac{d}{dt}(\mathrm{v}_{(y'')}) + (\mathrm{u}\omega\mathrm{r})\right]; \\ \dot{\mathbf{v}}_{(z'')} &= \mathbf{k}'' \frac{d}{dt}(\mathrm{v}_{(z'')}). \end{aligned}\right\} \qquad (228)$$

Hence, in order that these values may be reconciled with a permanent zero value for ($\mathbf{v}_{(y'')}$), the magnitude of (**u**) must be adjusted to the acceleration parallel to (**j**″) of the epoch, which for the present purpose we may suppose to be one among the data, as well as the velocity component (i″(ωr)). At the same time, as the forms of the last equation show, the accelerations parallel to $(\mathbf{k}'', \mathbf{i}'')$ are reckoned as though those were constant unit-vectors. But it is plain that the existence of shift cannot disappear completely from acceleration and from velocity too, because the necessary conditions

$$(\mathbf{u} \times \mathbf{r}) = 0; \qquad (\mathbf{u} \times \mathbf{v}) = 0; \qquad \text{with } (\mathbf{u}) \text{ not zero}; \tag{229}$$

are incompatible, so long as $(\mathbf{v})$ and $(\mathbf{r})$ are not parallel.

There is a strong natural suggestion, through the connection and the form in which these ideal coördinates have come to our attention, that they bear by their intention upon the astronomical problems that occupy themselves with orbits whose differential sectors are drawn out of one containing plane by disturbing forces. To this conception of a continuous succession of osculating orbits the method is ingeniously accommodated, with a separation that is of practical advantage between the forces $(m\dot{\mathbf{v}}_{(x'')})$, $(m\dot{\mathbf{v}}_{(z'')})$, whihc, as it were, control the orbit-element of the epoch, and the force $(m\dot{\mathbf{v}}_{(y'')})$ into which the distorting influence is collected. Yet interest in the method should not be confined to astronomers, because its device is repeated with only the modifications that the new conditions impose under the next heading, when the osculating circle of curvature is brought into relation with any curved path of a moving point; and the parallelism is an instructive feature for our discussion.

Resolution on Tangent and Normal.

112. The local tangent and normal to the path of a moving point afford a coördinate system that has been in general use since the days of Euler, but its employment for velocities could not be carried beyond the rudimentary stage of indicating the set of values (0, 0, v) in every such application. It is clear that this remark includes with equal force momentum and kinetic energy that contain no other kinematical factor than velocity.

The resolution tangentially and normally has that ground for concerning itself solely with acceleration and with dependent dynamical quantities like force, power and work. In this it differs from the coördinate systems that have been occupying us hitherto: by not being serviceable in more than one stage of differentiation, whereas the terms of the other systems have linked with two derivatives at least. How the tangent-normal plan branches off from the radius-vector series appears when we write

$$\mathbf{r} = \mathbf{r}_1 r; \qquad \dot{\mathbf{r}} \equiv \mathbf{v} = \dot{\mathbf{r}}_1 r + \mathbf{r}_1 \frac{dr}{dt}; \qquad \ddot{\mathbf{r}} \equiv \dot{\mathbf{v}} = \dot{\mathbf{v}}_1 v + \mathbf{v}_1 \frac{dv}{dt}; \tag{230}$$

and compare the last equality, that realizes the separation along tangent and normal to the path, with equation (209) that continues the polar component scheme. Because one stage does isolate itself thus, it becomes feasible for it to remain bound by the invariance test for a quantity with which it connects, and yet take on the quality of a mixed plan in other respects. A plan mixed or composite in regard to the standard frame, by dealing with comparison-frames (O′, X′Y′Z′) whose ($\mathbf{r}'$) and ($\dot{\mathbf{r}}'$) are not invariant with ($\mathbf{r}$) and ($\dot{\mathbf{r}}$), though ($\ddot{\mathbf{r}}'$) and ($\ddot{\mathbf{r}}$) are thus related. Section 77 furnishes all needed reminder about like combinations.

Such realities as the exclusion of normal force from effect upon power have thrown tangential force into stronger relief; and the more impressive function of the latter in changing magnitudes. Some plan or other of resolution for acceleration is favored, because the resultant quantity finds in general no visible geometrical element falling in its line as the tangent to the path does with velocity. The projections on tangent and normal form the simplest set that contains any segregation, for as we have once noticed, the (XYZ) set does not discriminate but speaks always of its own tensors. The separation on the basis that tangential acceleration changes velocity through its tensor alone, and the normal part changes the unit-vector alone, is the most important early and familiar instance that general ideas of vector algebra had to pattern after. The last of equations (230) has, as we are aware, grown into a general handling of any vector derivative.

113. The polar components of acceleration have been found to involve in comparative complexity the distinctive traits of the velocity vector as exhibited by its derivative, because their formulation is guided by elements foreign to (**v**) and borrowed from the behavior of the other vector (**r**). And as we see illustrated repeatedly, the changes in any vector indicate themselves most directly by analysis of its derivative according to some leading idea inherent in the vector itself. It did not escape us that the vector (**H**), for example, is but indirectly described by use of (ω) and ($\dot{\omega}$) in sections 56 and 57, and that there is likely to be a gain when the more direct connection of ($\dot{\mathbf{H}}$) and (**M**) is utilized.

Before entering upon any new considerations, let us once more pick up the thread at section 89, and renew the thought that (**xyz**) and (**x'y'z'**) can be read as projections of any free vector such as (**v**). Then equations (154) or their alternatives made explicit for ($\dot{x}$, $\dot{y}$, $\dot{z}$) are the algebraic statement of shift for acceleration, (**v**) for the standard frame being given indifferently by

$$\mathbf{v} = \mathbf{i}x + \mathbf{j}y + \mathbf{k}z; \qquad \mathbf{v} = \mathbf{i}'x' + \mathbf{j}'y' + \mathbf{k}'z'. \tag{231}$$

Also the details worked out for (**r**), beginning with section 78, are translatable for (**v**), and justify for instance, as we can use now the Euler angles, and are paralleling ($\mathbf{r}_0 = 0$),

$$d\mathbf{v} = \frac{\partial \mathbf{v}}{\partial t}\,dt + \frac{\partial \mathbf{v}}{\partial \psi}\,d\psi + \frac{\partial \mathbf{v}}{\partial \vartheta}\,d\vartheta + \frac{\partial \mathbf{v}}{\partial \varphi}\,d\varphi, \tag{232}$$

whose meaning reproduced more briefly in

$$\dot{\mathbf{v}} = \dot{\mathbf{v}}_{(m)} + (\dot{\boldsymbol{\gamma}} \times \mathbf{v}) \tag{233}$$

gives foundation for our next useful conclusion.

114. In a plane curve that is the path of a point (Q), the successive orientations of the tangent can be said to arise by a continuous turning, whose axis is the normal at (Q) to the plane of the path. And this turning to which we assign the angular velocity ($\boldsymbol{\omega}$), and which accompanies the progress with velocity ($\mathbf{v}$) of (Q) along the curve, is registered in its effect upon ($\mathbf{v}$) through the normal acceleration that is written

$$\dot{\mathbf{v}}_{(n)} = (\boldsymbol{\omega} \times \mathbf{v}) = -\boldsymbol{\rho}_1 \frac{v^2}{\rho}. \tag{234}$$

The order of factors in the second member is seen to direct this acceleration toward the local center of curvature of the path, and the known geometry introduces the radius of curvature, whose standard unit-vector points away from that center. Complementary to this is the tensor change in ($\mathbf{v}$) provided for by the tangential acceleration whose natural form is

$$\dot{\mathbf{v}}_{(t)} = \mathbf{v}_1 \frac{dv}{dt}. \tag{235}$$

In order to recast these statements in the language of shift, let comparison-frames be conceived distributed along the path and with origins in it, each in a permanent configuration with the standard frame, its (X′) axis pointing forward along the local tangent and its (Y′) axis inward along the normal, ($\boldsymbol{\omega}$) being standard as positive. All such frames will give both velocity and acceleration invariantly with the standard frame, and for each one as (Q) passes its origin the same conditions prevail at the epoch:

$$\mathbf{v}_{(x')} = \mathbf{v}; \qquad \mathbf{v}_{(y')} = \mathbf{v}_{(z')} = 0. \tag{236}$$

But the shift of origin alone, as we have noticed elsewhere, being without effect upon the projections as vectors, the application of equation (233) will yield

$$\dot{\mathbf{v}}_{(x')} = \mathbf{i}'\frac{d}{dt}(v_{(x')}) = \mathbf{v}_1\frac{dv}{dt}; \qquad (237)$$
$$\dot{\mathbf{v}}_{(y')} = \mathbf{j}'(\omega v); \qquad \dot{\mathbf{v}}_{(z')} = 0;$$

consistently with equations (234, 235).

115. But a space curve differs from a plane curve very much as the instantaneous orbit spoken of in section 111 differs from a plane orbit, in that its differential sectors, bounded now by radii of curvature and not by radius-vectors, are not coördinated into one plane. Each is treated typically like the uniplanar case, however, but in the plane of its epoch. A gradual change of this plane can always be accomplished by an added turning about some axis contained in each plane element, the displacements due to which being normal to that element are merely superposed on whatever process is being completed within the plane of the element itself. The direction of each such axis in its individual plane will be chosen according to the particular condition that it is desired to fulfil.

In the account of Hansen's coördinates it was proved that the designated axis left both component velocities $(\boldsymbol{\omega} \times \mathbf{r})$ and $(\mathbf{r}_1(dr/dt))$ unaffected by a rotation about it; and also two of the three component accelerations. In the example before us now, it becomes desirable to leave unchanged the one velocity $(\mathbf{v})$ that enters unresolved, and the entire acceleration. It soon appears how this is attained by letting each differential sector turn about an axis that is the line of $(\mathbf{v})$ at the epoch. This will add no new velocity at any point like (Q) in that axis, and it leaves the acceleration components unaltered because the supplementary term $(\dot{\boldsymbol{\gamma}}' \times \mathbf{v})$ would in any event be normal to the plane element, if $(\dot{\boldsymbol{\gamma}}')$ denotes a rate of rotation about any axis in that plane, and this term vanishes for every magnitude of $(\dot{\boldsymbol{\gamma}}')$

when the latter is colinear with ($\mathbf{v}$). Consequently if we apply equation (233) again, writing

$$\dot{\gamma} = (\omega + \dot{\gamma}'), \tag{238}$$

equations (236, 237) are continued in validity for any space curve, though derived originally from uniplanar motion. It is plain in what way the shift process is to be modified when it must include a varying plane (X′Y′) for the osculating circle; and also that the tensor of ($\dot{\gamma}'$) must be fitted to the *tortuosity* of the curve, while (ω) is determined by the circle of curvature. The vector magnitude ($\dot{\gamma}'$) is, to the extent shown, external to the acceleration problem stated; and in this it goes beyond the corresponding vector ($\mathbf{u}$) of Hansen's system, as reference to equation (228) confirms. The geometry of space curves, in which our axis (Z′) figures as the binormal, is seen to build with similar ideas to those just developed.

116. If a comparison-frame (O′, X′Y′Z′) is moving as a whole relative to the standard with unaccelerated translation whose velocity is ($\mathbf{v}_0$), and the velocity of (Q) relative to (O′, X′Y′Z′) is ($\mathbf{v}'$), the last of equations (230) gives for

$$\mathbf{v} = \mathbf{v}_0 + \mathbf{v}', \qquad \dot{\mathbf{v}} = \dot{\mathbf{v}}_1'\mathrm{v}' + \mathbf{v}_1'\frac{d\mathrm{v}'}{dt}. \tag{239}$$

And since by supposition ($\mathbf{i}'\mathbf{j}'\mathbf{k}'$) are here constant vectors, there is no distinction between ($\dot{\mathbf{v}}_1'$) relative to (X′Y′Z′) and (XYZ). Hence comparing the paths of (Q) relative to the two frames, it is clear that the sum is invariant, if we add together each tangential acceleration and its partner of normal acceleration, although the velocities in the paths are different, as is the apportionment of the acceleration between the two components. Such indifference as exists to the inclusion or the exclusion of constant velocities is often a helpful fact in treating of accelerations. But its other limitations must be observed beside the one just indicated, as applying for example to power ($\mathbf{R}\cdot\mathbf{v}$). If in this product (**R**) is retained, and (**v**) is changed to (**v**′), the product is altered unless ($\mathbf{v}_0$) and (**R**) happen to be perpendicular.

As the summation

$$\int_{s_0}^{s} ds = (s - s_0) \tag{240}$$

constitutes a rectification of the path, so the other legitimate summation

$$\int_{v_0}^{v} \left(m \frac{dv}{dt} dt \right) = m(v - v_0) \tag{241}$$

might be termed a rectification of momentum. In each operation we may see, by one way of viewing it, the accumulation of tensor elements upon one shifted line that becomes parallel in succession to the vector elements whose tensors are thus summed. But it does not explain fully why the second summation is mathematically as valid as the first, just to remark that each element of momentum is colinear with an element (d**s**). The tensor factors may be in any ratio that varies from one element to another and distorts the graph. In addition to whatever else can be said, we may return to the idea of comprehensive tensor running through a process of shift and observe what condition makes an element of actual displacement and the exact differential of such a tensor equal, by obviating that foreshortening of each element and the telescoping of their series that shift in general causes. If we take for instance equation (122) in connection with its context, the condition is seen to be that the vector product denoted generally by ($\dot{\gamma} \times \mathbf{V}$) should be perpendicular to the line on which the tensor in question is laid off. This becomes a specially simplified relation when the plan of shift is such that only one tensor occurs. The polar scheme contains only the length (**r**) of the radius-vector; the tangent and normal resolution only the tensor of (**v**); which may indeed be identified with (**r**) by the thought of section 88. In forming the derivative of (**r**) or of (**v**) under the form of equation (137), ($\dot{\mathrm{V}}_{(m)}$) comprises nothing but the total derivative of the tensor, and the mathematical test for integrability is met. If it were practically easier to devise plans of the type instanced, without sacrificing other advantages, there would be less hindrance to forming integrated values of tensors in working out results of shift.

117. We shall close this summary of our last system of point coördinates by gathering for record its most serviceable relations to the fundamental quantities, and here again with a representative particle at the center of mass of a body definitely in mind. They show in terms of projections parallel to the (X′Y′Z′) specified for equations (236, 237), with (x_0', y_0', z_0') added for the coördinates in the standard frame of the particle caught in passage through the (O′) of the epoch.

$$\left.\begin{aligned}
&\mathbf{Q}_{(x')} = \mathbf{Q} = m\mathbf{v}; \qquad \mathbf{Q}_{(y')} = \mathbf{Q}_{(z')} = 0;\\
&E_{(x')} = E = \tfrac{1}{2}mv^2; \qquad E_{(y')} = E_{(z')} = 0;\\
&\mathbf{H} = (\mathbf{x}_0' + \mathbf{y}_0' + \mathbf{z}_0') \times \mathbf{Q} = +\mathbf{j}'(z_0'Q) - \mathbf{k}'(y_0'Q);\\
&\mathbf{R}_{(x')} = \mathbf{i}'\left(m\frac{dv}{dt}\right); \qquad \mathbf{R}_{(y')} = \mathbf{j}'(mv\omega); \qquad \mathbf{R}_{(z')} = 0;\\
&P = R_{(x')}v = m\frac{dv}{dt}v = \frac{d}{dt}(E_{(x')});\\
&\mathbf{M} = (\mathbf{x}_0' + \mathbf{y}_0' + \mathbf{z}_0') \times (\mathbf{R}_{(x')} + \mathbf{R}_{(y')})\\
&\qquad = \mathbf{k}'\left(m\omega v x_0' - m\frac{dv}{dt}y_0'\right)\\
&\qquad + \mathbf{j}'\left(m\frac{dv}{dt}z_0'\right) - \mathbf{i}'(m\omega v z_0').
\end{aligned}\right\} \qquad (242)$$

The expression written for (**M**) should be compared with the direct vector derivative of (**H**) as given above in terms of the shifting (X′Y′Z′).

EULER'S DYNAMICAL EQUATIONS.

118. The configuration angles (ψ, θ, ϕ) have been associated with Euler's name already; and once more we follow the established custom in speaking of the next plan to be examined as Euler's, describing the statements of it as his dynamical equations, and so contrasting them with the purely geometrical or kinematical ideas brought forward under the other title.[1] This second group of Euler's equations constitutes a system of resolution for the dynamical quantities that departs in one important respect from all the others that have preceded it in the order that we are following. It has been constructed with specific reference to a rigid body as a whole, instead of being shaped for one element of mass, or at most for a particle at the center of mass. The summation covering the entire mass has been incorporated into the expressions, as an integral part of their standard form; the field of use for them is particularly among those parts of the total quantities that must fall outside all plans that are limited, in conception or in effective and convenient adaptation, to a particle's translation. Therefore it will be anticipated that we shall deal in these equations with that element of rotation in the most general type of motion for a rigid body, which is the obligatory remainder after deducting a translation with its center of mass. The explanations on this point in sections 48 to 63 may be referred to; also those in regard to the dynamical independence of the rotation and the translation, and the connection of a pure rotation about an origin with one about a moving center of mass (see sections 52 and 53). Let it be remarked, in order to cover this aspect of the situation, that Euler's dynamical equations once developed for the conditions of rotation, are applicable equally to either occurrence of it.

119. A junction with previous results can be made by bringing together the equations for the values of (**H**) and of (**M**), since

[1] See Note 28.

it has been proved that moment of momentum and force-moment furnish central clews to guide inquiry among the phenomena of rotation. Let the understanding be that our analysis attaches primarily to rotation about a center of mass (C′), and that any necessary transitions to pure rotation are to be adequately indicated.

On returning to equations (86) the signs of mass-summation are in evidence, and also of the general interrelation between each component of (**H**) and all three components of the rotation-vector (ω), when an unguided choice of (XYZ) has been made, to which axes those located at (C′) will be assumed parallel for a beginning. The concept of (ω) as properly applicable to the complex of radius-vectors lying within the body has been adopted profitably, but it is not to be overlooked that a changing configuration of body and (XYZ) makes the inertia factors variable. Neither does parallelism of the axis of (ω) with one of (XYZ), permanent or transient, introduce the lacking symmetry into these equations. Note, however, the form of equation (80), regard (ω) as parallel to (Z), and complete the set of component equations thus particularized. They are for (X′Y′Z′) at (C′),

$$\mathbf{H}_{(z')} = \mathbf{k}'(\omega_{(z')}I_{(z')}); \qquad \mathbf{H}_{(y')} = \mathbf{j}'(-\ \omega_{(z')}\textstyle\int_m y'z'dm); \qquad \mathbf{H}_{(x')} = \mathbf{i}'(-\ \omega_{(z')}\textstyle\int_m z'x'dm). \tag{243}$$

Observe the form of the last two components, and the fact that the orienting factors in them are coördinates.

120. The commentary of the last paragraph can be duplicated essentially in respect to equations (89), replacing (**H**) by (**M**′) and (ω) by ($\dot{\omega}$). Thus if we next suppose the axis of ($\dot{\omega}$) parallel to (Z), all three components of (**M**′) persist, and a similar difference in type reappears, between the first component and the two others. Again for (X′Y′Z′) at (C′),

$$\mathbf{M}'_{(z')} = \mathbf{k}'(\dot{\omega}_{(z')}I_{(z')}); \qquad \mathbf{M}'_{(y')} = \mathbf{j}'(-\ \dot{\omega}_{(z')}\textstyle\int_m y'z'dm); \qquad \mathbf{M}'_{(x')} = \mathbf{i}'(-\ \dot{\omega}_{(z')}\textstyle\int_m z'x'dm). \tag{244}$$

Bringing in the other part ($\mathbf{M}''$) of the total force-moment does not better the symmetry, neither of the last equations nor of their parent equations, since in reliance upon equations (75, 76) we find

$$\mathbf{M}'' = (\boldsymbol{\omega} \times \mathbf{H}). \tag{245}$$

These observations multiply reasons for appropriating the principal axes at (C′) in a selective choice of (X′Y′Z′) for any one epoch, and then perpetuating whatever advantages are reaped, by introducing a shift that is so regulated that the same three lines of the body which are its principal axes for (C′) shall always be taken to mark or indicate the configuration of the fixed frame, in terms of whose projections or components of the quantities in question the equations are to be written. The case for these principal axes is strengthened when equation (88) adds kinetic energy to the expressions in this way simplified; and when we reflect that within the scheme now proposed, the inertia factors are reduced from six in number to three that are the principal moments of inertia, and that the triplet retains the same values as the axes under this scheme shift. The general case is to be supposed, where there are no more than three principal axes at (C′), and the momental ellipsoid is not one of rotation.

In view of the rôle about to be assigned to them, a specialized notation referring to principal axes is called for, and we shall meet that need first by using (A, B, C) to denote both the magnitudes of the principal moments of inertia and the axes with which they are associated. As magnitudes, (A, B, C) are scalar factors in equations. They are associated with lines and not with either one direction in those lines, so they are not vector tensors. As axes for specifying configuration, (ABC) designate by convention one direction in each line. The cycle order is as they stand written, so that in the zero of configuration, (A) is parallel to (X), (B) to (Y), and (C) to (Z). The axis of

(C) is then (Z′) of our preceding notation, and it has at any epoch the angular coördinates (ψ, ϑ). The third angular displacement (ϕ) is about the (C) axis itself. (See section 93.) Secondly, projections of any vector upon the principal axes will be denoted as illustrated for (ω) and ($\dot{\omega}$) thus:

$$\omega = \omega_{(a)} + \omega_{(b)} + \omega_{(c)}; \qquad \dot{\omega} = \dot{\omega}_{(a)} + \dot{\omega}_{(b)} + \dot{\omega}_{(c)}; \tag{246}$$

and the corresponding unit-vectors by ($\mathbf{a}_1$, $\mathbf{b}_1$, $\mathbf{c}_1$).

Utilizing this notation, the equations brought under review above are reduced to the forms

$$\left.\begin{aligned} \mathbf{H} &= \omega_{(a)}A + \omega_{(b)}B + \omega_{(c)}C; \\ \mathbf{M}' &= \dot{\omega}_{(a)}A + \dot{\omega}_{(b)}B + \dot{\omega}_{(c)}C; \end{aligned}\right. \tag{247}$$

$$\begin{aligned} \mathbf{M}'' = \mathbf{a}_1(\omega_{(b)}\omega_{(c)}C &- \omega_{(c)}\omega_{(b)}B) \\ &+ \mathbf{b}_1(\omega_{(c)}\omega_{(a)}A - \omega_{(a)}\omega_{(c)}C) \\ &\qquad + \mathbf{c}_1(\omega_{(a)}\omega_{(b)}B - \omega_{(b)}\omega_{(a)}A); \end{aligned} \tag{248}$$

$$E = \tfrac{1}{2}[A\omega^2_{(a)} + B\omega^2_{(b)} + C\omega^2_{(c)}]. \tag{249}$$

And this yields for the similar components of the total moment ($\mathbf{M}$)

$$\left.\begin{aligned} \mathbf{M}_{(a)} &= \mathbf{a}_1[\dot{\omega}_{(a)}A + \omega_{(b)}\omega_{(c)}(C - B)]; \\ \mathbf{M}_{(b)} &= \mathbf{b}_1[\dot{\omega}_{(b)}B + \omega_{(c)}\omega_{(a)}(A - C)]; \\ \mathbf{M}_{(c)} &= \mathbf{c}_1[\dot{\omega}_{(c)}C + \omega_{(a)}\omega_{(b)}(B - A)]. \end{aligned}\right\} \tag{250}$$

The sequence of ideas by which these specialized equations have been reached should be attentively scrutinized, also the interpretation of the combinations at this stage. Equations (250) are evidently valid at any one epoch, and can be evaluated if these elements are known at that epoch:

(1) The orientation of the axes (ABC) in (XYZ), and the magnitudes (A, B, C);

(2) The vector ($\dot{\omega}$) in tensor and orientation;

(3) The vector (ω) in tensor and orientation.

121. In order to supply some other profitable details, and to put another link in the connections of these equations with general forms, we shall recur to equations (86) and differentiate with regard to time, the first of them for a sample. It is fundamental that the result must represent the projection of (**M**) upon (X), the latter being taken arbitrarily; and that with base-point at (C′) all moments must be reducible to couples, all net force being absorbed into the translation. (See section 51.) The conspicuous complication in this derivative is a lesson about what principal axes avoid, for we find

$$\dot{\mathbf{H}}_{(x)} = \mathbf{M}_{(x)} = \mathrm{i}\left\{ I_{(x)}\frac{d}{dt}(\omega_{(x)}) + \omega_{(x)}\frac{d}{dt}(I_{(x)}) - \omega_{(y)}\int_m \frac{dx}{dt}\,y\,dm - \omega_{(y)}\int_m x\frac{dy}{dt}\,dm - \frac{d}{dt}(\omega_{(y)})\int_m xy\,dm - \omega_{(z)}\int_m \frac{dx}{dt}\,z\,dm - \omega_{(z)}\int_m x\frac{dz}{dt}\,dm - \frac{d}{dt}(\omega_{(z)})\int_m zx\,dm \right\}. \quad (251)$$

In the third member, the third, fourth, sixth and seventh terms are to be further expanded by use of the velocity relations for rotation,

$$\frac{dx}{dt} = \omega_{(y)}z - \omega_{(z)}y; \qquad \frac{dy}{dt} = \omega_{(z)}x - \omega_{(x)}z; \qquad \frac{dz}{dt} = \omega_{(x)}y - \omega_{(y)}x. \quad (252)$$

When the axes (XYZ) are particularly chosen to be the set (ABC) in its position at the epoch, all terms can be struck out that contain as factors the integrals known as products of inertia. And this choice cancels the second term in the third member also. Because for all sets of orthogonal axes at the same origin we have

$$I_{(x)} + I_{(y)} + I_{(z)} = 2\int_m r^2 dm \text{ (an invariant magnitude)}; \quad (253)$$

and hence during relative displacement of body and (XYZ),

$$\frac{d}{dt}(I_{(x)}) + \frac{d}{dt}(I_{(y)}) + \frac{d}{dt}(I_{(z)}) = 0. \quad (254)$$

But for the longest and for the shortest axis of the momental ellipsoid, corresponding to the least and the greatest principal moment of inertia, the condition of maximum or minimum removes two terms separately from the above equation of condition, which then proves that a stationary value of moment of inertia enters for the third principal axis also.

After removing all the terms of indicated zero value, there remains

$$\dot{\mathbf{H}}_{(x)} = \mathbf{a}_1 \left\{ I_{(x)} \frac{d}{dt}(\omega_{(x)}) + \omega_{(y)}\omega_{(z)}\int_m (y^2 + x^2) dm - \omega_{(z)}\omega_{(y)}\int_m (z^2 + x^2) dm \right\}, \quad (255)$$

for comparison with the first of equations (250). The two statements harmonize completely, if we insist upon the identity of meaning for the expressions

$$\dot{\omega}_{(a)}, \qquad \mathbf{i}\left[\frac{d}{dt}(\omega_{(x)})\right]; \qquad \text{[(A) and (X) parallel.]}$$

they are both representative of the projection of the vector ($\dot{\omega}$) upon (A) or (X). The comparison for the two other pairs of equations is to be made similarly.

122. The next step in progress releases equations (250) from this one reading of their symbolism, and lays a foundation for the equivalences

$$\dot{\omega}_{(a)} = \mathbf{a}_1 \frac{d}{dt}(\omega_{(a)}); \qquad \dot{\omega}_{(b)} = \mathbf{b}_1 \frac{d}{dt}(\omega_{(b)});$$
$$\dot{\omega}_{(c)} = \mathbf{c}_1 \frac{d}{dt}(\omega_{(c)}), \quad (256)$$

where the second members are to be recognized as components

of $(\dot{\mathbf{V}}_{(m)})$ in equation (137), for application to the derivative $(\dot{\boldsymbol{\omega}})$ as expressed under a process whose shift rate is marked by the axes (ABC). Since these are definite lines of the body, they must conform to its rotation-vector $(\boldsymbol{\omega})$, and we have in this shift another example of cancelled correction, for

$$\dot{\boldsymbol{\omega}} = \dot{\boldsymbol{\omega}}_{(m)} + (\boldsymbol{\omega} \times \boldsymbol{\omega}) = \mathbf{a}_1 \frac{d}{dt}(\omega_{(a)}) + \mathbf{b}_1 \frac{d}{dt}(\omega_{(b)}) + \mathbf{c}_1 \frac{d}{dt}(\omega_{(c)}), \quad (257)$$

where the tensors in the third member have taken on a new shade of interpretation. They have become the generalized values for the shifting axes, instead of being particularized single values.

But there is one more consequence in this direction that still remains to be formulated, and that can be drawn from the expression in equations (247) for moment of momentum which can now be conceived as continuously valid and differentiated, due allowance being included for the changing orientation of the projections that make up the total. We can write

$$\dot{\mathbf{H}} = \left[\mathbf{a}_1 A \frac{d}{dt}(\omega_{(a)}) + \mathbf{b}_1 B \frac{d}{dt}(\omega_{(b)}) + \mathbf{c}_1 C \frac{d}{dt}(\omega_{(c)}) \right] + (\boldsymbol{\omega} \times \mathbf{H}), \quad (258)$$

whose separation into components restates equations (250), after incorporating into the latter the transitions of equations (256). The forms derived by either line of procedure are Euler's dynamical equations, whose establishment with the means at their inventor's disposal must always be rated as a remarkable achievement. It is in addition moreover remarkable that the segregation according to the terms of equation (258), which is more nearly mathematical in its origin, is also a separation that splits the force-moment into parts with a plain and important difference of physical effect; and the beginning made in section 55 was with design selected in order to dwell upon that fortunate chance. A conclusive proof of the equations in very few lines can evidently be extracted from the material that has been discussed here with greater expansion; but a demonstration may become too brief to be effective for insight, in a matter that has wide general bearings, so the detail is probably not superfluous.

123. Among the uses of Euler's equations, the predominant type of rigid body whose rotation is to be investigated is likely to show a certain symmetry, whose representation in the momental ellipsoid gives equality to two axes of the latter. This must convert the general ellipsoid into one of rotation with a symmetry axis; the known consequence being that all perpendiculars to that symmetry axis at the center of the rotational ellipsoid become principal axes with equal moments of inertia. This combination arises if the rotating body itself, being homogeneous in material, has an axis of symmetry; and bodies designed for rapid revolution are usually turned in a lathe. But it is clear that a prism of square cross-section, as well as a circular cylinder, would manifest its symmetry in a momental ellipsoid of rotation. And Euler's dynamical equations, being concerned with distribution of mass only as recorded in principal moments of inertia, would not discriminate between the two cases, granted the magnitudes (A, B, C) are severally equal in them.

It is proposed next to reconsider equations (250) in the light of this possibility, designating (C) as the axis of symmetry of the momental ellipsoid for (C′), with the corollary that the magnitudes (A) and (B) are equal; their common value we can call (A). If now the axes of (A) and (B) are still definitely located as lines of the body, whose rotation-vector ($\dot{\gamma}$) is identical therefore with (ω) for the body, no essential change appears in the equations except dropping out the last term of the third. Especially equations (256) that are determined by the equality of

($\dot{\gamma}$) and (ω) are available as before. However all lines of exposition in reaching Euler's equations must set the adoption of principal axes in the central place, and not the equality of the rotation-vectors. So by multiplying the number of principal axes the condition of symmetry enables choice to be variously exercised and yet range among them, though the auxiliary equality be abandoned and a relative motion through the material of the rotating body be permitted to the principal axes that have been selected. It is clear that the assumed relations limit the difference between ($\dot{\gamma}$) and (ω) to a turning about ($\mathbf{c}_1$) that is also ($\boldsymbol{\phi}_1$); but to this element it remains free to assign any magnitude. The expression of that freedom is

$$\begin{aligned} \omega &= \boldsymbol{\psi}_1 \frac{d\psi}{dt} + \boldsymbol{\vartheta}_1 \frac{d\vartheta}{dt} + \boldsymbol{\phi}_1 \frac{d\varphi}{dt}\,; \\ \dot{\gamma} &= \boldsymbol{\psi}_1 \frac{d\psi}{dt} + \boldsymbol{\vartheta}_1 \frac{d\vartheta}{dt} + \boldsymbol{\phi}_1 \left(k \frac{d\varphi}{dt}\right); \end{aligned} \tag{259}$$

where (k) may have any positive or negative value. Euler's equations proper given for ($k = 1$) have been put before us already; and we shall add for consideration, among the generalized Euler forms suggested by the last equation, only that modification which becomes necessary when the value of (k) is taken at zero. This supposition happens to offer some special advantage in handling combinations like a gyroscope under control by weight moment, and the earth as affected by a gravitation couple due to its spheroidal figure.

124. Let us mark the change of plan by using (A′, B′) to denote those principal axes that are now substituted for (A, B), recollecting first that as moments of inertia all four magnitudes are equal, and secondly, that (C) is common to both sets of axes. Then as a reminder of the needed revision in equations (256) we can write

$$\dot{\boldsymbol{\omega}} = \mathbf{a}_1{}' \frac{d}{dt}(\omega_{(a')}) + \mathbf{b}_1{}' \frac{d}{dt}(\omega_{(b')}) + \mathbf{c}_1 \frac{d}{dt}(\omega_{(c)})$$

$$+ \left(\left(\boldsymbol{\psi}_1 \frac{d\psi}{dt} + \boldsymbol{\vartheta}_1 \frac{d\vartheta}{dt} \right) \times \boldsymbol{\omega} \right). \quad (260)$$

In equations (193) are recorded values for those components of ($\dot{\boldsymbol{\omega}}$) which accord in directions with the present specifications for (A′B′C); and in equations (191) of section 101 the line of development caused us to put down in terms of ($\boldsymbol{\psi}$, $\boldsymbol{\vartheta}$, $\boldsymbol{\varphi}$) the first three entries on the right hand of equation (260). It seems advisable to clinch the comparison in respect to equations (256, 257) by developing here for that resolution the general components of ($\dot{\boldsymbol{\omega}}$), and lastly confirming the harmony of the two sets at their coincidence that occurs for ($\boldsymbol{\varphi} = 0$). These are the first details:

$$\left.\begin{aligned}
\dot{\boldsymbol{\omega}}_{(a)} &= \mathbf{a}_1 \frac{d}{dt}(\omega_{(a)}) = \mathbf{a}_1 \left[\frac{d^2\vartheta}{dt^2} \cos\varphi - \frac{d\vartheta}{dt}\frac{d\varphi}{dt} \sin\varphi \right. \\
&\quad + \frac{d^2\psi}{dt^2} \sin\vartheta \sin\varphi + \frac{d\psi}{dt}\frac{d\vartheta}{dt} \cos\vartheta \sin\varphi \\
&\quad \left. + \frac{d\psi}{dt}\frac{d\varphi}{dt} \sin\vartheta \cos\varphi \right]; \\
\dot{\boldsymbol{\omega}}_{(b)} &= \mathbf{b}_1 \frac{d}{dt}(\omega_{(b)}) = \mathbf{b}_1 \left[-\frac{d^2\vartheta}{dt^2} \sin\varphi - \frac{d\vartheta}{dt}\frac{d\varphi}{dt} \cos\varphi \right. \\
&\quad + \frac{d^2\psi}{dt^2} \sin\vartheta \cos\varphi + \frac{d\psi}{dt}\frac{d\vartheta}{dt} \cos\vartheta \cos\varphi \\
&\quad \left. - \frac{d\psi}{dt}\frac{d\varphi}{dt} \sin\vartheta \sin\varphi \right]; \\
\dot{\boldsymbol{\omega}}_{(c)} &= \mathbf{c}_1 \frac{d}{dt}(\omega_{(c)}) = \mathbf{c}_1 \left[\frac{d^2\varphi}{dt^2} + \frac{d^2\psi}{dt^2} \cos\vartheta \right. \\
&\quad \left. - \frac{d\psi}{dt}\frac{d\vartheta}{dt} \sin\vartheta \right];
\end{aligned}\right\} \quad (261)$$

the values to be differentiated in the second members being duly identified in the survey that equation (181) has put together. What remains of the results last written when they are particularized for the condition ($\dot{\boldsymbol{\phi}} = 0$); with ($\boldsymbol{\omega}_{(a')}$), ($\boldsymbol{\omega}_{(b')}$) obtained by a corresponding resolution of ($\boldsymbol{\omega}$), fills out the more general form of Euler's equations,

$$\left.\begin{aligned} \mathbf{M}_{(a')} &= \mathbf{a}_1'[\dot{\omega}_{(a')}A' + \omega_{(b')}\omega_{(c)}(C - B')]; \\ \mathbf{M}_{(b')} &= \mathbf{b}_1'[\dot{\omega}_{(b')}B' + \omega_{(c)}\omega_{(a')}(A' - C)]; \\ \mathbf{M}_{(c)} &= \mathbf{c}_1[\dot{\omega}_{(c)}C]; \end{aligned}\right\} \tag{262}$$

the necessitated companion being the equalities of magnitude

$$A = A' = B = B'. \tag{263}$$

Finally the components of ($\boldsymbol{\omega}$) that match the above statement being added:

$$\left.\begin{aligned} \boldsymbol{\omega}_{(a')} &= \mathbf{a}_1'\left(\frac{d\vartheta}{dt}\right); \qquad \boldsymbol{\omega}_{(b')} = \mathbf{b}_1'\left(\frac{d\psi}{dt}\sin\vartheta\right); \\ \boldsymbol{\omega}_{(c)} &= \mathbf{c}_1\left(\frac{d\varphi}{dt} + \frac{d\psi}{dt}\cos\vartheta\right); \end{aligned}\right\} \tag{264}$$

such advantage as this alternative formulation possesses on the kinematical side is made to appear. Dynamically something is contributed to a preference for it when the resultant force-moment is a vector that lies continually in the line of the axis (A′). A preliminary examination of the instances quoted above shows that they lend themselves unconstrainedly to this analysis which will be found applied in section 127.

125. On the surface the constant reference to ($\boldsymbol{\omega}$) and ($\dot{\boldsymbol{\omega}}$), either in their totals or through differently designated sets of their components, is apt to leave a misleading impression that they are pivotal quantities in any investigation where Euler's equations are employed. It seems worth while, therefore, to put in stronger light the primary emphasis of equation (258) upon changes that are going on in the moment of momentum vector (**H**). The separation in the second member there fits a line of demarcation between changes in magnitude and in direction, since the first group of terms is by the connections that have been established for it a magnitude derivative of

$$\mathbf{H} = \mathbf{a}_1(A\omega_{(a)}) + \mathbf{b}_1(A\omega_{(b)}) + \mathbf{c}_1(C\omega_{(c)}), \qquad (265)$$

though distorted from its value as reckoned in the standard frame by shift of the axes (ABC). But just that shift is indispensable, as we have insisted, in order that the properties of principal axes may prune the cumbrous algebraic expansions into maximum brevity. Where a corrected segregation for ($\dot{\mathbf{H}}$) into changes of magnitude and of direction entails a sacrifice of the gain by using (ABC), the balance of choice leans always one way; that much of dynamical indirectness in Euler's equations is condoned. But there is an increasing tendency and a wholesome one, to put their dynamical sense to the front, letting ($\boldsymbol{\omega}$) and ($\dot{\boldsymbol{\omega}}$) fall into a subordinate importance, derived in large degree from the clews they furnish to (**M**) and to the course of events for (**H**). It was less easy to do this under the older forms of Euler's day, but it is facilitated, as has perhaps been made convincingly apparent, by a vector algebra that follows so intimately the history of vector quantities.

126. Naturally the thought has suggested itself to inquire after a scheme modeled upon the resolution of force into a tangential and a normal component, for application to moment of momentum. One main obstacle is not difficult to detect, for after indicating the start in parallel to the other procedure,

$$\mathbf{H} = \mathbf{h}_1(H); \qquad \dot{\mathbf{H}} = \dot{\mathbf{h}}_1(H) + \mathbf{h}_1\left(\frac{dH}{dt}\right); \qquad (266)$$

it is noticeable first that (**H**) cannot be assumed to fall in a principal axis, and secondly that no data for ($\dot{\mathbf{h}}_1$) are available from geometrical sources. Therefore the longer forms, for (**H**) in equation (86) and for (dH/dt) in equation (251) must be used, and the expressions must be encumbered with an added angular velocity for ($\mathbf{h}_1$). Introduction of (XYZ) gives no help, nor of the partial time-derivatives that rely upon holding (ABC) stationary. Either leaves commingled the parts that are sought distinct.

But one resolution of force-moment can be carried through that is different from Euler's and yet has aspects that recommend it. This is contrived so that one component is taken in the axis of ($\boldsymbol{\omega}$) at each epoch, and arranged otherwise as will be explained presently; approaching in plan the tangential resolution of force in so far as ($\boldsymbol{\omega}$) and ($\mathbf{v}$) can be said to bear similar relations to the two aims. It has the merit besides of piecing out the usual discussion of rotation about a fixed axis, by giving recognition to those supplementary terms which disappear on fixing the axis about which the body is rotating.

Return to the value of ($\mathbf{M}''$) in equation (75) and of ($\mathbf{M}'$) formed by mass-summation of equation (82), and assemble their respective contributions. Let ($\mathbf{u}$) denote the rate of change in direction of ($\boldsymbol{\omega}$), so that with unit-vector ($\boldsymbol{\omega}_1$) we have

$$\dot{\boldsymbol{\omega}} = \boldsymbol{\omega}_1\left(\frac{d\omega}{dt}\right) + (\mathbf{u} \times \boldsymbol{\omega}); \tag{267}$$

where ($\mathbf{u}$) must be perpendicular to ($\boldsymbol{\omega}$); and subdivide ($\mathbf{r}$) as shown by

$$\mathbf{r} = \mathbf{r}_{(\omega)} + \mathbf{r}'. \tag{268}$$

Then

$$\mathbf{M}_{(\omega)} = \boldsymbol{\omega}_1\left[\frac{d\omega}{dt}\int_m(\mathbf{r}\cdot\mathbf{r})dm - \int_m r_{(\omega)}(\dot{\boldsymbol{\omega}}\cdot\mathbf{r})dm\right], \tag{269}$$

$$\dot{\boldsymbol{\omega}}\cdot\mathbf{r} = \left(\boldsymbol{\omega}_1\frac{d\omega}{dt} + (\mathbf{u}\times\boldsymbol{\omega})\right)\cdot(\mathbf{r}_{(\omega)} + \mathbf{r}')$$

$$= \left(\frac{d\omega}{dt}\,r_{(\omega)}\right) + (\mathbf{u}\times\boldsymbol{\omega})\cdot\mathbf{r}'. \tag{270}$$

Identify (Z′) with ($\boldsymbol{\omega}$), and (**u**) with (Y′) in direction, giving

$$\mathbf{M}_{(\omega)} = \mathbf{M}_{(z')} = \boldsymbol{\omega}_1 \left[\frac{d\omega}{dt} \int_m (r^2 - z'^2) dm - u\omega \int_m z'x'dm \right]$$

$$= \boldsymbol{\omega}_1 \left[\frac{d\omega}{dt} I_{(z')} - u\omega \int_m z'x'dm \right]. \qquad (271a)$$

In the plane (X′Y′) we have to consider

$$- \int_m (\boldsymbol{\omega} \times \mathbf{r})(\boldsymbol{\omega} \cdot \mathbf{r}) dm - \int_m \mathbf{r}'(\dot{\boldsymbol{\omega}} \cdot \mathbf{r}) dm + (\mathbf{u} \times \boldsymbol{\omega}) \int_m (\mathbf{r} \cdot \mathbf{r}) dm, \qquad (272)$$

from which are gathered without difficulty

$$\left.\begin{aligned} \mathbf{M}_{(x')} &= i' \left[\omega^2 \int_m y'z'dm - \frac{d\omega}{dt} \int_m z'x'dm + u\omega I_{(x')} \right], \\ \mathbf{M}_{(y')} &= j' \left[- \omega^2 \int_m z'x'dm - \frac{d\omega}{dt} \int_m y'z'dm \right. \\ &\qquad \left. - u\omega \int_m x'y'dm \right]. \end{aligned}\right\} \qquad (271b)$$

Noteworthy is the extent to which equations (271) are reduced by symmetries, though (**u**) is not zero, as well as the reappearance of the elementary form when (**u**) vanishes. Dissection of these moments shows almost immediately the force elements at (dm) in components parallel to our (X′Y′Z′) to be

$$\left.\begin{aligned} d\mathbf{R}_{(x')} &= i' \left[- \frac{d\omega}{dt} y' - \omega^2 x' \right] dm; \\ d\mathbf{R}_{(y')} &= j' \left[\frac{d\omega}{dt} x' - u\omega z' - \omega^2 y' \right] dm; \\ d\mathbf{R}_{(z')} &= \boldsymbol{\omega}_1 (u\omega y') dm; \end{aligned}\right\} \qquad (273)$$

which should be connected also with equation (72) by direct projection upon (X′Y′Z′), and by applying the proper shift process to ($\dot{\mathbf{H}}$), determined by the elements ($\boldsymbol{\omega}$, **u**) on the same line as sections 111 and 115 develop.

Regular Precession and Rotational Stability.

127. The aim and scope of these discussions could not attempt to include many particular requirements of individual problems without transgressing the boundary set by their intention, which is guided rather toward preparation for more generic or recurrent needs. It is, therefore, only because the dynamical features of gyroscopic action are generally acknowledged to be typical within a comparatively broad range, that some space is conceded to examination of them. But though this carries us beyond the stage of laying out a plan and somewhat into execution of it, it is proposed not to go far in that direction, nor to speak of more than two topics that are critical points in the general perspective. The first of these takes the form of a deliberate inquiry into the circumstances of that adjustment to steadied motion which is described with a phrase of wide acceptance as *regular precession,* and about which as a center so much else can be made to figure as a disturbance of it or a departure from it. And the second is devoted to laying bare the play of dynamical factors that operates to produce *rotational stability.*[1]

The arrangement of the gyroscope is assumed to give it a pure rotation about a fixed point (O), that is now taken as origin for axes like (A′B′C), the last named being an axis of symmetry, the shift rate for the set being as agreed in section 124, and the zero of configuration being marked by coincidence of (A′B′C) with (XYZ), where the (Z) axis is chosen vertical and downwards. The total controlling force-moment is supposed to be furnished by weight, the standard frame being fixed relatively to the earth, and the gyroscope has *universal joint* freedom at (O). For its rotation-vector (ω) then, the two equivalents have been supplied,

[1] See Note 29.

$$\omega = \psi_1 \frac{d\psi}{dt} + \vartheta_1 \frac{d\vartheta}{dt} + \phi_1 \frac{d\varphi}{dt} = \mathbf{a}_1' \left(\frac{d\vartheta}{dt} \right)$$

$$+ \mathbf{b}_1' \left(\frac{d\psi}{dt} \sin \vartheta \right) + \mathbf{c}_1 \left(\frac{d\varphi}{dt} + \frac{d\psi}{dt} \cos \vartheta \right). \quad (274)$$

For regular precession the conditions that obtain are

$$\frac{d\vartheta}{dt} = 0; \qquad \frac{d\psi}{dt}, \frac{d\varphi}{dt}, \vartheta, \text{ constant}; \qquad \text{or} \qquad (275)$$

$$\omega_{(a')} = 0; \qquad \omega_{(b')}, \omega_{(c)}, \text{ constant.}$$

And in order to standardize values, attach the further conditions

$$\frac{d\varphi}{dt} > 0; \qquad 0 < \vartheta < \frac{\pi}{2}; \qquad A' > C. \qquad (276)$$

Then the weight moment is negatively directed in the axis (A′), and with understandable notation the application of equations (262) to this adjustment shows the following scheme of specialized values:

$$\left.\begin{aligned} \mathbf{a}_1'(-W\bar{r} \sin \vartheta) &= \mathbf{a}_1' \left[A \frac{d\psi}{dt} \frac{d\varphi}{dt} \sin \vartheta \right. \\ &\quad \left. + (C - A) \left(\frac{d\psi}{dt} \sin \vartheta \right) \left(\frac{d\varphi}{dt} + \frac{d\psi}{dt} \cos \vartheta \right) \right]; \\ 0 &= \mathbf{b}_1' \, [\text{zero}]; \\ 0 &= \mathbf{c}_1 \, [\text{zero}]. \end{aligned} \right\} \quad (277)$$

It is a clear matter of algebra that the first equation is satisfied for

$$\sin \vartheta = 0;$$

or for

$$\frac{d\psi}{dt} = \frac{\omega_{(c)} C \pm \sqrt{(C\omega_{(c)})^2 + 4W\bar{r}A \cos \vartheta}}{2A \cos \vartheta};$$

or in another expression of it for

$$\frac{d\psi}{dt} = \frac{C \frac{d\varphi}{dt} \pm \sqrt{\left(C \frac{d\varphi}{dt} \right)^2 + 4W\bar{r}(A - C) \cos \vartheta}}{2(A - C) \cos \vartheta}. \quad (278)$$

Putting aside for the moment the first root, our questioning begins with ascertaining the dynamical double process that finds expression in the two signs of the second root and that shows to inspection in either form under the assumed relations of value, a quicker rotation about (Z) and a slower rotation of opposite sign as possible adjustments.

128. It lies on the surface that while regular precession continues the vector (**H**) can be changing its orientation only and not its tensor, and that since (**H**) must always be contained in the plane (B′, C), the applied force-moment must in the adjustment meet the condition

$$\mathbf{a}_1'(-\,W\bar{r}\sin\vartheta) = \left(\psi_1 \frac{d\psi}{dt} \times \mathbf{H}\right) \qquad (279)$$

equally at the quicker rate and at the slower rate of rotation about the vertical axis. For the explanation how this can occur, we shall look upon the moment of momentum as built up by superposition, following the second member of equation (274) in its elements which are now the first and third only. The contribution from the principal axis (C) and its horizontal part effective here in (**M**) let us write

$$\mathbf{H}_{(c)} = \mathbf{c}_1\left(C\frac{d\varphi}{dt}\right); \qquad \mathbf{N}' = \mathbf{n}_1\left(C\frac{d\varphi}{dt}\sin\vartheta\right). \qquad (280)$$

Then having excluded (Z) from being a principal axis by the suppositions laid down in the inequalities (276), the second instalment of (**H**) must allow for both a vertical and a horizontal part, the latter being contained in the plane (Z, C); and it alone is effective in (**M**); call it (**N″**). The total effective component of (**H**) for the vector product of equation (279) is accordingly an algebraic sum

$$\mathbf{N}' + \mathbf{N}'' = \mathbf{n}_1\left[\pm C\frac{d\varphi}{dt}\sin\vartheta \pm (A - C)\frac{d\psi}{dt}\sin\vartheta\cos\vartheta\right], \qquad (281)$$

the part (**N″**) being readily evaluated to confirm this.

129. It is next apparent from the cycle order that the rotation about (Z) must be negative in order that both terms within the parenthesis may first point the same way relatively to ($\mathbf{n}_1$) for our fixed assumptions, and secondly, give by the vector product that negative orientation in (A′) which the operative and negative weight-moment demands. So the standardized form in the circumstances becomes

$$\mathbf{M}_{(a')} = \left(\boldsymbol{\psi}_1 \frac{d\psi}{dt} \times \mathbf{H} \right)$$
$$= \mathbf{a}_1 \left[C \frac{d\psi}{dt} \frac{d\varphi}{dt} \sin \vartheta - (A - C) \left(\frac{d\psi}{dt} \right)^2 \sin \vartheta \cos \vartheta \right]. \qquad (282)$$

It is patent how elastic the constancy of this algebraic sum can be made, or of its equivalent vector product; large ($\mathbf{N}' + \mathbf{N}''$) and slow rotation, or smaller ($\mathbf{N}' + \mathbf{N}''$) and quicker rotation. With equation (281) besides to show reversal of the rotation about (Z) converting a numerical sum into an algebraic one, all other elements being held unchanged. But leaving those details as covered sufficiently, it behooves us to note in equations (278) that each double value has its own common quantities that are not entirely reconcilable. Since

$$\boldsymbol{\omega}_{(c)} = \boldsymbol{\varphi}_1 \left(\frac{d\varphi}{dt} + \frac{d\psi}{dt} \cos \vartheta \right), \qquad (283)$$

the first member, together with both (ϑ) and ($d\varphi/dt$), cannot all remain unchanged while the rotation about (Z) is made fast or slow. Equation (281) has tacitly taken one choice; but ($\boldsymbol{\omega}_{(c)}$) is a standard-frame quantity, whose constancy in magnitude moreover is assured under the third of equations (262) whenever ($\mathbf{M}_{(c)}$) is zero. We might then attach our thought preferably to the first form in equations (278), and recast the result thus:

$$\mathbf{M}_{(a')} = \left(\boldsymbol{\psi}_1 \frac{d\psi}{dt} \times \mathbf{H} \right) = \boldsymbol{\psi}_1 \frac{d\psi}{dt} \times (\mathbf{N}' + \mathbf{N}'')$$
$$= \mathbf{a}_1 \left[\left(\frac{d\psi}{dt} \sin \vartheta \right) \left(C\omega_{(c)} - A \frac{d\psi}{dt} \cos \vartheta \right) \right]; \qquad (284)$$

in which the possibilities of varying factors in a constant product reappear, with (ϑ) and ($\omega_{(c)}$) barred from change. It will be noticed finally that either more direct derivation of result corresponds exactly with the terms to which the first of equations (277) reduces, so our analysis reversed could be applied immediately to the latter. It ought to be said about the realization of conditions, that the spin round the (C) axis is usually preponderant heavily in magnitude, and for this reason the observed rotation about the vertical with a negative weight moment is normally *retrograde,* the necessary high rate for the contrary rotation being practically unattained.

130. Let equations (262) next be released from their restriction to that adjustment whose relations are now ascertained. Then with repetition of the idea put forward in the connection of section 56 there can be a rearrangement in this instance, too, that will describe the general action in terms of a deviation from adjustment as a convenient basis for exhibiting the consequences in the light of a disturbance. Re-establishing their unspecialized character, equations (277) will be written

$$\left.\begin{aligned}
\mathbf{a}_1'(-\mathrm{W}\bar{\mathrm{r}}\sin\vartheta) &= \mathbf{a}_1'\left[\mathrm{A}\frac{d^2\vartheta}{dt^2} + \frac{d\psi}{dt}\sin\vartheta\,\mathrm{C}\omega_{(c)}\right.\\
&\qquad\left. - \frac{d\psi}{dt}\cos\vartheta\,\mathrm{A}\frac{d\psi}{dt}\sin\vartheta\right];\\
0 &= \mathbf{b}_1'\left[\mathrm{A}\frac{d}{dt}(\omega_{(b')})\right.\\
&\qquad\left. + \frac{d\psi}{dt}\cos\vartheta\,\mathrm{A}\frac{d\vartheta}{dt} - \frac{d\vartheta}{dt}\mathrm{C}\omega_{(c)}\right];\\
0 &= \mathbf{c}_1\left[\mathrm{C}\frac{d}{dt}(\omega_{(c)})\right.\\
&\qquad\left. + \frac{d\vartheta}{dt}\mathrm{A}\frac{d\psi}{dt}\sin\vartheta - \frac{d\psi}{dt}\sin\vartheta\,\mathrm{A}\frac{d\vartheta}{dt}\right].
\end{aligned}\right\} \quad (285)$$

But all the items there put down only elaborate still the one dynamical fact that no vector change in moment of momentum is ever being produced except the increment along the instantaneous position of the (A′) axis, which is that of ($\boldsymbol{\vartheta}_1$). Denote the projection of (**H**) upon the plane (Z, C) by (**H′**), and the first of the three expressions can be put in these equivalent forms:

$$\left.\begin{aligned} \mathbf{a}_1'(-\,W\bar{r}\sin\vartheta) &= \mathbf{a}_1'\left[\frac{d}{dt}\left(A\frac{d\vartheta}{dt}\right)\right] + \left(\boldsymbol{\psi}_1\frac{d\psi}{dt}\times\mathbf{H}'\right); \\ \mathbf{M} - \mathbf{M}_{(0)} &= \mathbf{a}_1'\left[\frac{d}{dt}\left(A\frac{d\vartheta}{dt}\right)\right]. \end{aligned}\right\} \quad (286)$$

The first statement is read that the weight moment devotes to changing magnitude for the component of (**H**) in its own line whatever margin remains after providing for continuance of change in direction for the rest of (**H**). And the second, that the deviation of the actual moment (**M**) from the adjustment moment ($\mathbf{M}_{(0)}$) required for prevailing values is registered in a process of change for ($\boldsymbol{\vartheta}$). The indicated preëmption claim of the changes in direction has a certain figurative shading, we may allow, but a certain truth also; because those affect quantities at their existent values for the epoch, whereas the quantities that are changes in magnitude are called into being and not present already. And so with the second form of statement: the section referred to concedes that the subtracted force-moment in the first member may be declared nominal or mathematical; but both points of view above are dynamically suggestive and to be entertained as a mental habit.

The other equations of the group (285) set forth the kinematical complications that ensue because nothing dynamical is effective in those lines. They give foundation for important and interesting studies that are, however, only to be alluded to here; we shall content ourselves with insisting once more upon the thought of sections 56 and 57. At the regular precession adjustment every term in the second members of these equations vanishes separately and they become a blank recording nothing. Now they sum up algebraically to zero, though the individual terms need not vanish; but they are, in a sense to be understood with due limitations, as empty of physical content as ever; they chronicle only formal and internal readjustments of expression.

131. The topic of rotational stability is also at its core dynamical, and it is approachable most directly through the considerations that we have been attaching to regular precession, when the possibilities are examined of securing that type of adjustment with the (C) axis directed nearly in the upward vertical. We shall confine inquiry, on this side as well, to outlining the connections; their essentials being grasped, the exhaustive treatment of details offers no other obstacles than the inevitable mathematical difficulties.

The first pertinent thought is derivable from equations (278) when a range into the second quadrant is permitted to ($\boldsymbol{\vartheta}$), and a discrimination needs to be regarded between real and imaginary values of the rotation about (Z), or between adjustments that can and that cannot be accomplished. Selecting the first alternative form for the solution, this dividing line is to be drawn where the values denoted here as special yield the relation

$$0 = (C\omega'_{(c)})^2 + 4W\bar{r}A \cos \vartheta'; \qquad \cos \vartheta' < 0. \qquad (287)$$

And the critical magnitude which ($\omega_{(c)}$) must at least reach if imaginary values are to be excluded completely is given by

$$\omega_{(c)} = \pm \sqrt{\frac{4W\bar{r}A}{C^2}}; \qquad (288)$$

so that if the spin about (C) equals or exceeds this rate, the attainment of regular precession at every position in ($\boldsymbol{\vartheta}$) is only a matter of providing the companion value of the spin about (Z). With this simple mathematics clear the next step is, as in the previous combination, to detect and assign the dynamical reason that must underlie it. The first stage in meeting that requirement starts with the merely reshaped equation

$$A\frac{d^2\vartheta}{dt^2} = \sin\vartheta\left(-W\bar{r} - \frac{d\psi}{dt}C\omega_{(c)} + A\left(\frac{d\psi}{dt}\right)^2\cos\vartheta\right). \quad (289)$$

This can be made to tell us that if the axis (C), having been directed vertically upwards, moves away from that position, and changes (ϑ) by a small amount from the value (π), it will be true that

$$\frac{d^2\vartheta}{dt^2} = \frac{\Delta\vartheta}{A}\left(W\bar{r} + \frac{d\psi}{dt}C\omega_{(c)} + A\left(\frac{d\psi}{dt}\right)^2\right);$$
$$\Delta\boldsymbol{\vartheta} = \boldsymbol{\vartheta}_1\left(\pm\frac{d\vartheta}{dt}\right)dt; \qquad \cos\vartheta = -1. \quad (290)$$

In words, the rotation rate ($d\vartheta/dt$) will always be subject to reduction in magnitude when the above parenthesis is itself a negative quantity; and we have discovered a cause for this reduction by seeing how the weight moment meets a first claim for guiding directional changes in (**H**); a special case under equation (286) is before us now. The stronger such absorption of force-moment, the more rapid becomes that check upon the initial motion in ($\boldsymbol{\vartheta}$), which will begin straightway as (C) leaves the upward vertical whenever the parenthesis is in the aggregate negative. Therefore we are led by these considerations to look at equation (284) in a somewhat new light after rewriting it

$$\frac{d^2\vartheta}{dt^2} = 0 = -\frac{\sin\vartheta}{A}\left(W\bar{r} + \frac{d\psi}{dt}C\omega_{(c)} - A\left(\frac{d\psi}{dt}\right)^2\cos\vartheta\right). \quad (284a)$$

Then a zero value of the parenthesis when its factor is not zero marks the transition between favorable and unfavorable conditions for checking an existing motion in ($\boldsymbol{\vartheta}$). In application to the second quadrant, the third term must be a positive magnitude always, but it decreases as (C) approaches a horizontal position. It is clear that cases may occur where the first member has unfavorable sign as (ϑ) leaves the value (π), and becomes favorable only after a finite drop of the axis (C). Also it has been seen that the unfavorable interval can then be narrowed by quickening the spin about (C), and it disappears at the critical value indicated by equation (288). Because ($\sin \vartheta = 0$) is always one solution, there is a discontinuity possible here between the two types of solution, similar to that for the conical pendulum obtainable by assuming ($d\varphi/dt$) zero in the second form of equation (278). The classification sometimes made of gyroscope tops as weak and strong follows the line of thought just traced.

132. The factors in the second term of the parenthesis that is under examination are never quite independent so long as ($d\psi/dt$) occurs in ($\omega_{(c)}$); but their dependence assumes a special phase when the (C) axis and the vertical can become coincident, for then there will be only two different expressions for the same (vertical) component of (**H**). In order to develop the latter relation and to reduce the parenthesis accordingly we shall begin with the more general statement and afterwards particularize it. By projecting from (B′) and from (C) on the vertical and adding we obtain

$$\mathbf{H}_{(\psi_1)} = \mathfrak{k}_1 \left[\left(\frac{d\psi}{dt} \sin \vartheta B' \right) \sin \vartheta + \left(\frac{d\varphi}{dt} + \frac{d\psi}{dt} \cos \vartheta \right) C \cos \vartheta \right]. \quad (291)$$

Consequently

$$\mathbf{H}_{(\psi_1)} - \mathfrak{k}_1(\mathbf{H}_{(c)} \cdot \mathfrak{k}_1) = \mathfrak{k}_1 \left(\frac{d\psi}{dt} A \sin^2 \vartheta \right); \qquad [B' = A]; \quad (292)$$

with the general value for the tensor ratio

$$\frac{d\psi}{dt} = \frac{H_{(\psi_1)} - H_{(c)} \cos \vartheta}{A(1 - \cos^2 \vartheta)}, \quad (293a)$$

which gives under the equality attendant upon coincidence in the upward vertical, the conventions for signs being duly reconciled,

$$\frac{d\psi}{dt} = -\frac{C\omega_{(c)}}{2A}. \qquad [\cos\vartheta = -1.] \qquad (293b)$$

Substitution in equation (290) shows as a condition that the right-hand member should be negative when (C) leaves the upward vertical with positive $(\Delta\vartheta)$

$$(C\omega_{(c)})^2 > 4AW\bar{r}. \qquad (294)$$

The greater this inequality the stronger the retardation, the sooner the departure is brought to a halt. The mathematics of equation (288) has found thus a foundation in the dynamical process initiated when (C) leaves its vertical position.

133. In what precedes, the emphasis falls upon moment of momentum in relation to force-moment. The thought is not complete however until the work of the weight moment has been connected with changes in kinetic energy. For the case in hand we find by using the principal axes,

$$E = \tfrac{1}{2}A\left[\left(\frac{d\vartheta}{dt}\right)^2 + \left(\frac{d\psi}{dt}\sin\vartheta\right)^2\right] + \tfrac{1}{2}C\omega^2_{(c)}; \qquad (295)$$

and the last term being constant, the variations or interchanges consequent upon work done are confined to the two other terms. Now referring to equations (285) examination soon convinces us that the initiative, so to speak, centers in the quantity that is in the line of the resultant force-moment. So long as $(d\vartheta/dt)$ is zero, no change can occur in $(\omega_{(b')})$; but the vanishing of $(\omega_{(b')}, \omega_{(c)})$ separately or simultaneously might not prevent changes in $(d\vartheta/dt)$. It is characteristic of the stability here in question that the action depends vitally upon the actual occurrence of a displacement; and this accounts for the known feature of gyroscopic mechanisms, that their efficiency is nullified by removing the *degree of freedom* upon which their functioning depends.

For the power as the derivative of the kinetic energy, we can write

$$P = A\left[\left(\frac{d\vartheta}{dt}\right)\frac{d^2\vartheta}{dt^2} + \omega_{(b')}\frac{d}{dt}(\omega_{(b')})\right] = M_{(a')}\frac{d\vartheta}{dt}. \quad (296)$$

Let the conditions be such that positive work is done, negative moment being accompanied by negative displacement. Then the first term in the second member will be negative for opposite signs of its factors. And we see diverted from their appearance in the coördinate (ϑ) the magnitude changes in both (**H**) and (E) that (**M**) would make visible there, were there no gyroscopic interactions.

The general agreement of the equation (288) and the inequality (294) in their formulation of a critical value is obvious; and it ought not to be longer obscure why the same truth is at the foundation of each criterion. The essence of the adjustment to regular precession is the insufficiency of the available weight moment at a certain value of (ϑ) and other quantities to do more than supply exactly what is needed for the corresponding directional change in (**H**). The reversal in sense of the inequality that we arrived at, declares in effect an unavoidable preponderance of weight moment consistently with the other given values, and its sufficiency to quicken the motion in ($\boldsymbol{\vartheta}$) that is supposed to exist already. It is an easily deduced consequence therefore as regards the axis (C) that it will continue its departure from the upward vertical until conditions alter. The imaginary range of equation (278) is one signal that the combination of the accompanying spin about (Z) with the actual horizontal component of (**H**) is within that region unequal to monopolizing the full force-moment active. The quantitative elaboration of these leading ideas produces the accepted results in every detail.

Generalized Momenta and Forces.

134. At the date of their original announcement, Lagrange's coördinates and the equations of motion that employed them were contrived in the service of what would now be called mechanics proper, for the imperious reason that the longer list of energy transformations which dynamics distinctively embraces had not yet been discovered and drawn into the fundamental quantitative connections. The terms coördinate, configuration, velocity and momentum were enlarged by Lagrange from usage as he found it no doubt, but his broader scheme did not break the alliance with geometrical ideas for its kinematics. His parameters were ultimately based on combinations of lengths and position angles, though kept unspecialized by suppressing or deferring the analysis of them into the plainer geometrical elements. The energy too was introduced primarily in its kinetic form, that and momentum deriving their dynamical quality from those inertia factors that are in their nature either directly given as mass, or else as literal as moments of inertia that emerge from a mass-summation.[1]

Lagrange's equations will be found akin to Euler's in two respects: first they are normally intended for treating as a unit some body or system of bodies; and secondly, they are after a fashion of their own indifferent toward a substitution of one system for another, provided that determinate equivalencies are observed, as we have seen Euler's equations to be under invariance of (A, B, C) in magnitude. This likeness extends far enough to coördinate the two plans and to make the latter when duly stated a special result of Lagrange's broader handling. The demonstration offered by Lagrange himself is founded on d'Alembert's principle; and this interconnection of the two phases of the same idea, and of each with Hamilton's different formulation of it, lends to the establishment of the equations of motion an air of logical redundancy. This was the subject of a passing remark in our Introduction; and it might be recalled too that the noticeable swing away from the first vogue of d'Alem-

[1] See Note 30.

bert's statement centers upon a recent discovery of more comprehensive adaptability in the alternative forms devised by Lagrange and by Hamilton to a range of energy transformations that was unsuspected when either of the latter was first accepted. By the light of what is developing further in that quarter the estimate of their fruitfulness will continue to be decided.

Because these are the origins it seems advisable to let the treatment here conform to them, instead of making a short path to the newest reading. There is ground to expect that the fuller realization of meaning in the extension of method and of its valid possibilities will have its best source in a reasoned appreciation of where the latent power resided and how it was implanted. We hold one reliable clew already, wherever it proves true that a *mechanism*, construing the word not too remotely from direct perceptions, can be seen to give in its fluxes of energy and momentum a quantitative equivalent for those fluxes under less restricted conditions of transformation.

135. On working outwards to occupy a broader field, and passing at points the limits earlier drawn, some elements of new definition or specification are involved, which the circumstances lead toward supplying in part positively, in part by noting the barriers that remain. And we shall relinquish the attempt to finish each topic in a systematic progressive order, wherever it promises better success to proceed less rigidly; coming back to add a stroke and explain or define what was at first only sketched.

When it is said that any set of coördinates must determine a configuration completely, the plain idea is that they do for a system what we expect of the standard frame (O, XYZ), the coördinates being enumerated for as many joints or articulations as removal of ambiguity makes necessary. If the coördinate set is thus equivalent to (xyz), the same idea may be conveyed by declaring each general coördinate to be a definite function of the set (xyz). In normal usage we do not abandon the relation upheld for other coördinate systems, that the values expressed with their aid are standard frame values of the quantities dealt with, but we seek that aid through any convenient functions of (xyz) and not merely through lines and angles. Such preliminary conception of a coördinate denoted by (κ) prepares the way for a definition of the corresponding velocity as ($\dot{\kappa}$), meaning the total time-derivative of the magnitude of (κ), the question about vector quality being left open, an equal number of velocities and of coördinates being matched each to each.

Passing next to momentum we are again confronted with a definition that pairs each velocity with its own momentum quantity. Let (q) denote one of these momenta belonging to the velocity ($\dot{\kappa}$); then the defining equation is written, if (E) is still the total kinetic energy of the system to be studied,

$$q \equiv \frac{\partial E}{\partial \dot{\kappa}}. \tag{297}$$

And another fixed point in the scheme now being presented is that (E) shall be a homogeneous quadratic function of all the velocities ($\dot{\kappa}$). To this specification other things must be made to bend should that become necessary, which is a matter for due inquiry. But meanwhile one evident consequence of it can be read from the last equation, regarding the constitution of the momenta (q); they cannot be other than linear functions of the velocities ($\dot{\kappa}$) and homogeneous. Refer however to the closing remark of section 141.

136. Putting together what has been said, one feature in the relation of coördinates to configuration is caused to stand in relief: they must determine it in a form free from all reference to velocities in order that (E) may take on the assigned type. Let us add as being naturally required, that the members of a coördinate set must be mutually independent, and proceed to speak of their connection with the so-called *degrees of freedom* that a system of bodies possesses. Consideration of simplest instances, like that of a ball carried on the last in a numerous set of rods jointed together, shows that a large number of specifying elements or coördinates may be actually employed in designating configuration, even in one plane. But we know also that two rectangular or two polar coördinates only are required in this case; and the prevailing distinction seems to follow the line thus indicated, making degrees of freedom equal in number to the minimum group of coördinates requisite in describing a configuration, classing the excess in the number really used as *superfluous coördinates*. This disposes of the matter well enough, leaving for special examination only such interlocking of two coördinates into related changes as happens when a ball rolls (without sliding) on a table; and that finer point need not detain us. In these terms, a rigid solid has available not more than six degrees of freedom, three of which might call for coördinates locating its center of mass, with the remaining three covered by the Euler angles, for example. And we may borrow from regular procedure in that case, as known through repeated discussion, that an equation of motion is associated with each degree of freedom. That normal arrangement continues with evident good reason, though our treatment is shaped according to Lagrange's proposals, which do not change the objective in essence, but only the mode of reaching it.

137. To complete the plan, therefore, into which accelerations do not enter directly, there is need to specify its forces; here the determining thought has its root in the energy relations, running in the course that we shall next lay out, whose first stage has no novelty, but merely holds to the equivalence in work established for any resultant force. The right to substitute one force (**R**) for all the distributed effective force elements depends upon its equality with them in respect to total work and impulse. The same thought, in other words, declares equal capacity for setting up the total flux of kinetic energy and momentum in relation to the system of bodies, the separation of force and couple moment or of translation and rotation being a detail and without final influence. It is inherent, moreover, in the determination of any such resultant through vector sums or through algebraic sums that a set of components may be variously assigned to the same resultant. The ground that Lagrange traversed led him to a variation only on previous forms in expressing this essential energetic equivalence of the resultant force. The fact indeed that he set out from the *equilibrium principle of virtual work* due to d'Alembert should obviate any surprise on meeting the defining equation for his *generalized forces.*

With each degree of freedom which makes flux of kinetic energy possible, associate its force (F); sum the work during elements of displacement in all the coördinates (κ) and express its necessary equality to the same work given in terms of the usual forces parallel to (X, Y, Z). The equation is

$$\Sigma(\mathrm{Fd}\kappa) = \Sigma\int_{\mathrm{m}}(\mathrm{d}\mathbf{R}\cdot\mathrm{ds}) = \Sigma\int_{\mathrm{m}}[\mathrm{dR}_{(x)}\mathrm{dx} + \mathrm{dR}_{(y)}\mathrm{dy} + \mathrm{dR}_{(z)}\mathrm{dz}], \quad (298)$$

which yields by a transformation that embodies through the partial derivative notation the supposition of independence that goes with the coördinates, for each force an expression

$$\mathrm{F} = \Sigma\int_{\mathrm{m}}\left[\mathrm{dR}_{(x)}\frac{\partial \mathrm{x}}{\partial \kappa} + \mathrm{dR}_{(y)}\frac{\partial \mathrm{y}}{\partial \kappa} + \mathrm{dR}_{(z)}\frac{\partial \mathrm{z}}{\partial \kappa}\right]. \quad (299)$$

Holding to this statement any force (F) can be defined in magnitude by the work per unit of displacement in its coördinate; and the narrowing assumption does not appear that (F) and (κ) are colinear, provided a convention can be observed that gives the work its real sign as determined by gain or loss to the system's kinetic energy. It is this relation which Lagrange's equations enlarge by including the other energies of dynamics.

We continue by introducing necessarily equivalent expressions for a change in configuration,

$$dx = \Sigma\left(\frac{\partial x}{\partial \kappa}\, d\kappa\right); \qquad dy = \Sigma\left(\frac{\partial y}{\partial \kappa}\, d\kappa\right); \qquad dz = \Sigma\left(\frac{\partial z}{\partial \kappa}\, d\kappa\right); \tag{300}$$

in which the summation extends to all the coördinates (κ). Then in the fluxion notation

$$\dot{x} = \Sigma\left(\frac{\partial x}{\partial \kappa}\, \dot{\kappa}\right); \qquad \dot{y} = \Sigma\left(\frac{\partial y}{\partial \kappa}\, \dot{\kappa}\right); \qquad \dot{z} = \Sigma\left(\frac{\partial z}{\partial \kappa}\, \dot{\kappa}\right); \tag{301}$$

from which follow for each coördinate singly the important equalities

$$\frac{\partial \dot{x}}{\partial \dot{\kappa}} = \frac{\partial x}{\partial \kappa}; \qquad \frac{\partial \dot{y}}{\partial \dot{\kappa}} = \frac{\partial y}{\partial \kappa}; \qquad \frac{\partial \dot{z}}{\partial \dot{\kappa}} = \frac{\partial z}{\partial \kappa}. \tag{302}$$

Taking the term from the first integral of equation (299), it can be given the form, by using the last results

$$dR_{(x)} \frac{\partial x}{\partial \kappa} = \frac{d}{dt}\left(dQ_{(x)} \frac{\partial \dot{x}}{\partial \dot{\kappa}}\right) - dQ_{(x)} \frac{d}{dt}\left(\frac{\partial x}{\partial \kappa}\right); \tag{303a}$$

and similarly from the remaining integrals,

$$\left.\begin{aligned} dR_{(y)} \frac{\partial y}{\partial \kappa} &= \frac{d}{dt}\left(dQ_{(y)} \frac{\partial \dot{y}}{\partial \dot{\kappa}}\right) - dQ_{(y)} \frac{d}{dt}\left(\frac{\partial y}{\partial \kappa}\right); \\ dR_{(z)} \frac{\partial z}{\partial \kappa} &= \frac{d}{dt}\left(dQ_{(z)} \frac{\partial \dot{z}}{\partial \dot{\kappa}}\right) - dQ_{(z)} \frac{d}{dt}\left(\frac{\partial z}{\partial \kappa}\right). \end{aligned}\right\} \tag{303b}$$

To recast the last factors in these three equations we write

$$\frac{d}{dt}\left(\frac{\partial}{\partial \kappa}(x)\right) = \frac{\partial \dot{x}}{\partial \kappa}; \qquad \frac{d}{dt}\left(\frac{\partial}{\partial \kappa}(y)\right) = \frac{\partial \dot{y}}{\partial \kappa}; \qquad \frac{d}{dt}\left(\frac{\partial}{\partial \kappa}(z)\right) = \frac{\partial \dot{z}}{\partial \kappa}; \tag{304}$$

whose justification is somewhat a matter of mathematical conscience. The order of the two differentiations may boldly be inverted as a legitimate operation; or whatever hazard may be felt in that can be guarded against by rigorous proofs that are accessible. Incorporating the last forms and summing equations (303), the force finds expression as

$$F = \Sigma\int_m \left[\frac{d}{dt}\left(\frac{\partial}{\partial \dot{K}}(dE)\right) - \frac{\partial}{\partial K}(dE)\right]$$

$$= \frac{d}{dt}\left(\frac{\partial E}{\partial \dot{K}}\right) - \frac{\partial E}{\partial K}, \quad (305)$$

in application to each one of the coördinates, and the whole development is then open to further comment or illustration.

138. This exposition of Lagrange's equations, and of the concepts upon which their statement rests, has been kept apart purposely from the infusion of vectorial ideas, in order to set forth as clearly as may be done that possibility upon which their larger usefulness in great measure depends, and of which insistent mention was made in the first chapter. Some care seems needed to break up the misleading connotations of words like velocity and momentum, that in their first and perhaps most literal sense imply each an orienting vector. And the emancipation of thought in this regard has been hindered doubtless by the unsuggestive practice of pointing out as examples of this method of attack solely those where velocities and momenta and forces offer themselves habitually as vectors—like those which our material has been including hitherto. If the trend of any demonstration equivalent to the foregoing be watched, however, it is seen to hinge essentially upon an enumeration of a sum of terms in the total energy of all forms that are considered, and analyzing them as products that conform to a type. This contains always as a factor the time rate of one in a group of quantities by whose means the changes in that energy content are adequately determined. The success of the analysis therefore depends, broadly speaking, upon the isolation of suitable factors in the physics of the energy forms to specify the *energy configuration* and to provide the necessary *velocities*. And in that direction it is interesting to note the part really played by the (XYZ) velocities and momenta as they lead to the vital connections in equation (305). They are scarcely more than a scaffolding, an aid in building but removed from the structure built, impressing effectively only one character upon the result—that its scheme of values shall be quantitatively a possible set in that mechanical phantom or model which is mirrored in the case treated. On their face, Lagrange's equations might seem to stand in parallel with tangential ordinary forces only, since the latter are alone concerned in work. But we shall show that this limitation does not in fact exist, and that the pattern set by the (XYZ) axes when they include for their projections constraints as well, is stamped upon these other combinations, which may be caused also to reveal normal forces that may be active (see section 141). As a counterpart to this relation it is to be observed how the (XYZ) axes fit everywhere into a plan of algebraic products through their three coexistent and practically scalar operations; and how for the element of scalar mass equations (1, II) are always free alternatives, whatever restrictions subsequent steps may impose, as for instance equation (67) has recorded.

139. Having laid some preliminary emphasis upon the extent to which they may exceed in scope other coördinate systems, it will be advisable to carry the comparison with Lagrange's plans into the region of overlapping, and make this last system prove itself capable of bringing out correct consequences there too, when orientation is reëstablished. The cross relations have many lessons that are of value; and some are yielded by a review of the polar coördinates that we shall put first. Borrowing from section 106 the expression for kinetic energy of a particle, and using fluxion notation for brevity,

$$E = \tfrac{1}{2}m[\dot{r}^2 + r^2\dot{\vartheta}^2 + r^2 \sin^2 \vartheta(\dot{\psi})^2]. \tag{306}$$

The Lagrange coördinates must be independent and sufficient to give configuration in (XYZ); and ($\mathbf{r}$, $\boldsymbol{\vartheta}$, $\boldsymbol{\psi}$) meet this requirement. But the velocities must correspondingly be ($\dot{r}$, $\dot{\vartheta}$, $\dot{\psi}$). The details work out into the forms, $(\partial E/\partial \psi)$ being zero,

$$\left.\begin{aligned}
&\frac{\partial E}{\partial \dot{r}} = m\dot{r}; \quad \frac{\partial E}{\partial \dot{\vartheta}} = mr^2\dot{\vartheta}; \quad \frac{\partial E}{\partial \dot{\psi}} = mr^2 \sin^2 \vartheta\dot{\psi};\\
&\frac{d}{dt}\left(\frac{\partial E}{\partial \dot{r}}\right) = m\ddot{r}; \quad \frac{d}{dt}\left(\frac{\partial E}{\partial \dot{\vartheta}}\right) = m(2r\dot{\vartheta}\dot{r} + r^2\ddot{\vartheta});\\
&\frac{d}{dt}\left(\frac{\partial E}{\partial \dot{\psi}}\right) = m(2r \sin^2 \vartheta\dot{\psi}\dot{r} + 2r^2 \sin \vartheta \cos \vartheta\dot{\psi}\dot{\vartheta}\\
&\qquad\qquad + r^2 \sin^2 \vartheta\ddot{\psi});\\
&\frac{\partial E}{\partial r} = m(r\dot{\vartheta}^2 + r \sin^2 \vartheta(\dot{\psi}))^2; \quad \frac{\partial E}{\partial \vartheta} = m(r^2 \sin \vartheta \cos \vartheta(\dot{\psi}))^2.
\end{aligned}\right\} \tag{307}$$

A general agreement is at once manifest when these terms are grouped and compared with equations (208); but it is a striking difference that the forces ($F_{(\vartheta)}$) and ($F_{(\psi)}$), associated with those two coördinates, must now be recognized as moments of the forces denoted previously by ($\mathbf{R}_{(x')}$) and ($\mathbf{R}_{(y')}$), for rotation-axes characterized plainly through the respective lever arms. This is a necessary concomitant of making velocities out of ($\dot{\vartheta}$, $\dot{\psi}$). The regrouping of terms also is instructive in betraying that loss of distinction for the orientation changes here as well which algebra usually evinces.

140. For a second example, let us make in the Lagrange form a restatement of section 89, utilizing equations (154) as a starting-point, and adapting them to a particle, as the desirably simple case. If (x′, y′, z′) are selected as three coördinates, the configuration in (XYZ) is not determinate by them alone, but in the plan followed the position angles for the axes (X′Y′Z′) must be known also; and of these as many as are independent can be added to make the required list of coördinates, of which all but three will then be superfluous in a sense already explained, and not to be reckoned among the degrees of freedom. The purpose of illustration can be attained sufficiently if we consider the uniplanar conditions, both for the particle which is then supposed to be restricted to the (XY) plane, and for the relative configuration of (X′Y′Z′), where we assume (Z) and (Z′) permanently coincident. Hence for the kinetic energy of (m) the expression is in understandable terms

$$E = \tfrac{1}{2}(\dot{x}^2 + \dot{y}^2)m = \tfrac{1}{2}m[\dot{x}'^2 + \dot{y}'^2 + (x'^2 + y'^2)\dot{\gamma}^2 - 2\dot{x}'y'\dot{\gamma} + 2x'\dot{y}'\dot{\gamma}], \quad (308)$$

the coördinates being now (x', y', γ) and the velocities $(\dot{x}', \dot{y}', \dot{\gamma})$; the last velocity is an algebraic derivative, (Z) being the fixed axis for (γ). Again the details are, when this homogeneous quadratic function of the velocities is differentiated,

$$\left.\begin{aligned}
&\frac{\partial E}{\partial \dot{x}'} = m(\dot{x}' - y'\dot{\gamma}); \qquad \frac{\partial E}{\partial \dot{y}'} = m(\dot{y}' + x'\dot{\gamma});\\
&\frac{\partial E}{\partial \dot{\gamma}} = m(\dot{\gamma}(x'^2 + y'^2) - \dot{x}'y' + x'\dot{y}');\\
&\frac{\partial E}{\partial x'} = m(\dot{\gamma}^2 x' + \dot{\gamma}\dot{y}');\\
&\frac{\partial E}{\partial y'} = m(\dot{\gamma}^2 y' - \dot{\gamma}\dot{x}'); \qquad \frac{\partial E}{\partial \gamma} = 0.
\end{aligned}\right\} \quad (309)$$

After forming the time-derivatives of the first three in the group and substituting values, we obtain for the three forces of the coördinates,

$$\left.\begin{aligned}
F_{(x')} &= m(\ddot{x}' - \ddot{\gamma}y' - 2\dot{\gamma}\dot{y}' - \dot{\gamma}^2 x');\\
F_{(y')} &= m(\ddot{y}' + \ddot{\gamma}x' + 2\dot{\gamma}\dot{x}' - \dot{\gamma}^2 y');\\
F_{(\gamma)} &= x'F_{(y')} - y'F_{(x)'}.
\end{aligned}\right\} \quad (310)$$

The third coördinate advertises that it is superfluous, in that its force value, whose form is readily verifiable as a moment, only confirms what is otherwise ascertained about the remaining forces.

141. In their adaptation to the present class of cases, some truths can be picked out that furnish clews for the lines of more extended use. First, referring to equations (155) and collating them with equations (302, 304), the latter are seen to be far-reaching analogues of changes that build upon the line of the quantity at the epoch, and of those others that depend upon a change of slope; they are correlated respectively with changing tensor and orientation of a vector. While a partial derivative like $(\partial x/\partial \kappa)$ may appear as a direction cosine within the purely geometrical conditions, it is a more inclusive *reduction factor* elsewhere. It is also open to observation in the last two illustrations that the generalized momenta become for those applications the orthogonal projections upon a distinguishable line, either of the momentum or of the moment of momentum in the standard frame. Differences of distribution for the same total projection between various pairs of groups is no more than part of the mathematical machinery, and it is especially to be expected where sets of partial derivatives occur whose variables have been changed. Note that

$$\frac{\partial E}{\partial \dot{\kappa}}; \qquad \frac{\partial E}{\partial \kappa}; \tag{311}$$

presuppose: the first, that all coördinates are held stationary, and all velocities but that one; and the second, that only the one coördinate is allowed to change, and none of the velocities. Comparisons with other sets of partials in our developments should prove helpful, as it will be to find answer for the question whether the Lagrange plan, when it deals with forces like (**R**), affiliates more closely with the mode of equation (112) or with that of equation (233).

Related to the second example here and to the ideas about superfluous coördinates, is another point of view that has likeness with the method of section 82. The standard frame coördinates, as expressed in equations (150), can be discriminatingly dependent upon time, indirectly through (x′, y′, z′) and directly through the direction cosines. Their exact differentials will then appear as

$$dx = \frac{\partial x}{\partial x'} dx' + \frac{\partial x}{\partial y'} dy' + \frac{\partial x}{\partial z'} dz' + \frac{\partial x}{\partial t} dt \qquad (312)$$

with two companions, the last term in each comprising the group that arise by differentiating the direction cosines if we have regarded (xyz) as given in a functional form like

$$x = f(x', y', z', t), \qquad (313)$$

and the superfluous elements are spoken of and dealt with as due to *variations of the geometrical relations with time.* The distinction that such changes of direction are assigned and not brought about by physical action is consistent with what has been seen above—the absence of those additional force specifications that would be introduced through them otherwise. The exercise of preference in selecting the elements to be drawn off thus into their own time function, however, need not be always the plainest of matters. And where an accompanying verbal usage is accepted that denies the title coördinate to position variables not ranked among degrees of freedom, the kinetic energy ceases to be a homogeneous quadratic function of the (remaining) legalized velocities. Of course these comments hold good for extension to the generalized *energy configuration.*

142. Retaining the energy value and imposing upon equations (310) the conditions that ($\dot{\gamma}$) and the origin shall be so regulated as to keep ($\mathbf{v}_{(y'')}$) at zero permanently, they conform to the tangent and normal resolution of force for those uniplanar restrictions, and in space curves there is the same correspondence between the general case and the one duly specialized. The test of the latter form being of some length and of no difficulty, and because it shows finally only an equivalent for section 115, we pass it with mention merely and proceed to examine Euler's equations for instructive connections with those of Lagrange.

We can quote two equally valid expressions for rotational energy of a rigid solid for which (A = B), when mounted as in section 127:

$$E = \tfrac{1}{2}(\omega^2_{(a')} + \omega^2_{(b')})A + \tfrac{1}{2}\omega^2_{(c)}C$$
$$= \tfrac{1}{2}(\dot{\vartheta}^2 + (\dot{\psi}\sin\vartheta)^2)A + \tfrac{1}{2}(\dot{\varphi} + \dot{\psi}\cos\vartheta)^2C. \quad (314)$$

In the former, no total time-derivatives can be detected of quantities determining configuration, but only those projections of a given (ω) appear which presuppose knowledge of the configuration, and which could be rated partial derivatives of (γ) according to the explanation of equation (185) as related to section 79. This fact has been noticed in several connections since the subject of position angles was opened (see sections 93 and 98), and it explains why the direct expression by means of the Euler angles is not entirely superseded by using ($\omega_{(a)}$, $\omega_{(b)}$, $\omega_{(c)}$). The coördinates are then (ψ, ϑ, ϕ), the velocities ($\dot{\psi}$, $\dot{\vartheta}$, $\dot{\varphi}$) in the fluxion notation, and we foresee that our previous force-moments will now figure as forces. It is plain that

$$\frac{\partial E}{\partial \psi} = \frac{\partial E}{\partial \varphi} = 0; \qquad F_{(\psi)} = F_{(\phi)} = 0; \quad (315)$$

the latter pair of values expressing the controlling constancies of the moment of momentum in this problem, or of the momenta ($q_{(\psi)}$, $q_{(\phi)}$) in the present terminology. These values when worked out, and those that complete the expression

$$F_{(\vartheta)} = \frac{d}{dt}(q_{(\vartheta)}) - \frac{\partial E}{\partial \vartheta}, \quad (316)$$

are all in recognizable identity with what was obtained elsewhere.

143. The action of the gyroscope has been seen capable of diverting energy from one coördinate to another as a perhaps secondary consequence of maintaining change of direction in a moment of momentum that is of constant magnitude. And it is easy to multiply instances, wherever the inertia factors (moments and products of inertia) can be variable, that a change in value for kinetic energy is demanded under constancy of the other quantity, this being entailed if the rotation factor alters. Thus a symmetrically shrinking homogeneous sphere has constant (**H**) under the influence of gravitational self-attractive forces between its parts, but the rotational energy grows as an expression of work done in the shortening lines of stress. In symbols, for rotation about a diameter,

$$\mathbf{H} = \omega \mathrm{I}_{(\mathrm{D})}; \qquad \mathrm{E} = \tfrac{1}{2}\omega^2 \mathrm{I}_{(\mathrm{D})} = \frac{1}{2}\left(\frac{\mathrm{H}}{\mathrm{I}_{(\mathrm{D})}}\right)^2 \mathrm{I}_{(\mathrm{D})} = \frac{\mathrm{H}^2}{2\mathrm{I}_{(\mathrm{D})}}, \tag{317}$$

with the denominator growing continually smaller. What is here illustrated is more widely possible to happen among the analogous factors of energy, where its different forms are interconnected in the same system, so that the energy may be transferred and redistributed among the Lagrange coördinates though some of the corresponding momenta remain unaltered. Neither is it remote from the mental attitude already alluded to, in approaching the study of a physical system through certain external and accessible bearings of it while a margin is left for less definite inference, to base tentative conclusions about concealed constant momenta upon observable indirect effects on energy. It is some preparation for those fields of usefulness to follow out the relations in the next sequence of ideas, which may be carried through first for directed momenta and finally be restated more broadly.

We shall suppose a system with four generalized coördinates, three (ψ, ϑ, φ) what we have termed accessible, and details about the fourth (τ) to be subjects for inference, as we may say. The latter has then naturally no force assigned to it for direct connection with changes of energy, and is adapted to the thought expressed above, by having its momentum assumed a constant magnitude. Accordingly these conditions are written

$$F_{(\tau)} = 0; \qquad q_{(\tau)} = \text{constant}. \tag{318}$$

Add the supposition as conforming reasonably to the limitations upon knowledge, that no known relations contain (τ) itself. Then since

$$F_{(\tau)} = \frac{d}{dt}(q_{(\tau)}) - \frac{\partial E}{\partial \tau}, \tag{319}$$

each term in the second member vanishes separately or is a blank.

144. The momentum $(q_{(\tau)})$ being actually present can modify the phenomena; that is the effects of other forces and the energy reactions. It is to be asked: How will the statements be recast, if we detect $(q_{(\tau)})$ as though distributed in parts added to the other momenta, to which the phenomena are being exclusively ascribed? This moves in the direction of suspending direct inquiry into (τ), so the method is frequently described as allowing *ignoration of coördinates.*[1] Expressing this resolution of $(q_{(\tau)})$ with the aid of the direction cosines (l, m, n), and adding its components to the other momenta as indicated, the total orthogonal projections on the lines will indicate

$$\frac{\partial E}{\partial \dot{\psi}} = q'_{(\psi)} + l q_{(\tau)}; \qquad \frac{\partial E}{\partial \dot{\vartheta}} = q'_{(\vartheta)} + m q_{(\tau)};$$
$$\frac{\partial E}{\partial \dot{\varphi}} = q'_{(\phi)} + n q_{(\tau)}. \tag{320}$$

The coördinates (ψ, ϑ, ϕ) need not be themselves orthogonal, but the parts (q') and $(q_{(\tau)})$ are.

The adjudged energy (E) would then have to satisfy the general relation growing by implication out of the real scalar product for rotation

[1] See Note 31.

$$E = \tfrac{1}{2}(\omega \cdot \mathbf{H}), \tag{321}$$

the possible non-linearity of any velocity ($\dot{\kappa}$) and its momentum (q) being here also recognized; this yields the form

$$E = \tfrac{1}{2}[\dot{\psi}(q'_{(\psi)} + lq_{(\tau)}) + \dot{\vartheta}(q'_{(\vartheta)} + mq_{(\tau)}) + \dot{\varphi}(q'_{(\phi)} + nq_{(\tau)})]. \tag{322}$$

Introducing (Q) in this connection to denote the constant magnitude ($q_{(T)}$), the forces derivable from the supposed energy will appear as containing the terms

$$\left.\begin{aligned} \frac{d}{dt}\left(\frac{\partial E}{\partial \dot{\psi}}\right) &= \frac{d}{dt}\left(\frac{\partial E_{(0)}}{\partial \dot{\psi}}\right) + Q\frac{dl}{dt}; \\ \frac{d}{dt}\left(\frac{\partial E}{\partial \dot{\vartheta}}\right) &= \frac{d}{dt}\left(\frac{\partial E_{(0)}}{\partial \dot{\vartheta}}\right) + Q\frac{dm}{dt}; \\ \frac{d}{dt}\left(\frac{\partial E}{\partial \dot{\varphi}}\right) &= \frac{d}{dt}\left(\frac{\partial E_{(0)}}{\partial \dot{\varphi}}\right) + Q\frac{dn}{dt}. \end{aligned}\right\} \tag{323}$$

The quantity of energy ($E_{(0)}$) represents what would be present if (Q) were non-existent, and the last terms in the equations register the modification due to the introduction of (Q) on the supposed basis, namely through its resolved parts that maintain the directions of the momenta ($q_{(\psi)}, q_{(\vartheta)}, q_{(\phi)}$). Their indicated connection with changes of direction relative to (ψ, ϑ, φ) momenta should not pass unnoticed. To conform with the above values, the energy (E′) allowed for in excess of ($E_{(0)}$) must be

$$E' = \dot{\psi}(lQ) + \dot{\vartheta}(mQ) + \dot{\varphi}(nQ); \tag{324}$$

and in order to fill out consistently the scheme begun with equations (323) we must continue in the expressions of force with

$$\left.\begin{aligned} \frac{\partial E}{\partial \psi} = \frac{\partial E_{(0)}}{\partial \psi} + \frac{\partial E'}{\partial \psi}; \quad \frac{\partial E}{\partial \vartheta} = \frac{\partial E_{(0)}}{\partial \vartheta} + \frac{\partial E'}{\partial \vartheta}; \\ \frac{\partial E}{\partial \varphi} = \frac{\partial E_{(0)}}{\partial \varphi} + \frac{\partial E'}{\partial \varphi}. \end{aligned}\right. \tag{325}$$

But we find

$$\left.\begin{aligned}\frac{\partial E'}{\partial \psi} &= Q\left(\dot{\psi}\frac{\partial l}{\partial \psi} + \dot{\vartheta}\frac{\partial m}{\partial \psi} + \dot{\varphi}\frac{\partial n}{\partial \psi}\right);\\ \frac{\partial E'}{\partial \vartheta} &= Q\left(\dot{\psi}\frac{\partial l}{\partial \vartheta} + \dot{\vartheta}\frac{\partial m}{\partial \vartheta} + \dot{\varphi}\frac{\partial n}{\partial \vartheta}\right);\\ \frac{\partial E'}{\partial \varphi} &= Q\left(\dot{\psi}\frac{\partial l}{\partial \varphi} + \dot{\vartheta}\frac{\partial m}{\partial \varphi} + \dot{\varphi}\frac{\partial n}{\partial \varphi}\right).\end{aligned}\right\} \tag{326}$$

Hence the aggregate departures from the forces that would be indicated by ($E_{(0)}$) alone can be seen in

$$\left.\begin{aligned}\frac{d}{dt}\left(\frac{\partial E}{\partial \dot{\psi}}\right) - \frac{\partial E}{\partial \psi} &= \frac{d}{dt}\left(\frac{\partial E_{(0)}}{\partial \dot{\psi}}\right) - \frac{\partial E_{(0)}}{\partial \psi}\\ &\quad + Q\left[\frac{dl}{dt} - \left(\dot{\psi}\frac{\partial l}{\partial \psi} + \dot{\vartheta}\frac{\partial m}{\partial \psi} + \dot{\varphi}\frac{\partial n}{\partial \psi}\right)\right];\\ \frac{d}{dt}\left(\frac{\partial E}{\partial \dot{\vartheta}}\right) - \frac{\partial E}{\partial \vartheta} &= \frac{d}{dt}\left(\frac{\partial E_{(0)}}{\partial \dot{\vartheta}}\right) - \frac{\partial E_{(0)}}{\partial \vartheta}\\ &\quad + Q\left[\frac{dm}{dt} - \left(\dot{\psi}\frac{\partial l}{\partial \vartheta} + \dot{\vartheta}\frac{\partial m}{\partial \vartheta} + \dot{\varphi}\frac{\partial n}{\partial \vartheta}\right)\right];\\ \frac{d}{dt}\left(\frac{\partial E}{\partial \dot{\varphi}}\right) - \frac{\partial E}{\partial \varphi} &= \frac{d}{dt}\left(\frac{\partial E_{(0)}}{\partial \dot{\varphi}}\right) - \frac{\partial E_{(0)}}{\partial \varphi}\\ &\quad + Q\left[\frac{dn}{dt} - \left(\dot{\psi}\frac{\partial l}{\partial \varphi} + \dot{\vartheta}\frac{\partial m}{\partial \varphi} + \dot{\varphi}\frac{\partial n}{\partial \varphi}\right)\right].\end{aligned}\right\} \tag{327}$$

145. But the energy really introduced by the momentum (Q), like the other portion $E_{(0)}$ of the energy is expressible by a homogeneous quadratic function of the velocities which it is permissible at any one epoch to put into the form

$$E_{(Q)} = \tfrac{1}{2}K(\dot{\tau} + l\dot{\psi} + m\dot{\vartheta} + n\dot{\varphi})^2, \tag{328}$$

(K) being a function of coördinates only, and the value being in other respects fixed by necessary relations for partial derivatives of $E_{(Q)}$. Thus

$$\left.\begin{aligned} &\frac{\partial E_{(Q)}}{\partial \dot{\tau}} = K(\dot{\tau} + l\dot{\psi} + m\dot{\vartheta} + n\dot{\varphi}) = Q \text{ [by definition];} \\ &\frac{\partial E_{(Q)}}{\partial \dot{\psi}} = lQ; \quad \frac{\partial E_{(Q)}}{\partial \dot{\vartheta}} = mQ; \quad \frac{\partial E_{(Q)}}{\partial \dot{\varphi}} = nQ \\ &\qquad\qquad\qquad\qquad \text{[by equations (320)].} \end{aligned}\right\} \tag{329}$$

Further we have, since ($E_{(Q)}$) involves coördinates through both factors,

$$\frac{\partial E_{(Q)}}{\partial \psi} = \frac{1}{2}\frac{\partial K}{\partial \psi}(\dot{\tau} + l\dot{\psi} + m\dot{\vartheta} + n\dot{\varphi})^2 + K(\dot{\tau} + l\dot{\psi} + m\dot{\vartheta} + n\dot{\varphi})\left(\dot{\psi}\frac{\partial l}{\partial \psi} + \dot{\vartheta}\frac{\partial m}{\partial \psi} + \dot{\varphi}\frac{\partial n}{\partial \psi}\right); \tag{330}$$

and the second part is recognizable through equations (327, 329). In order to adapt the remaining part to the present connection, first put equation (328) into the legitimate form next shown, and then express its partial derivative for a coördinate, subject to our condition that (Q) is a constant magnitude. The results are

$$\begin{aligned} E_{(Q)} &= \frac{1}{2}\frac{Q^2}{K}; \\ \frac{\partial E_{(Q)}}{\partial \psi} &= -\frac{1}{2}\frac{Q^2}{K^2}\frac{\partial K}{\partial \psi} = -\frac{1}{2}\frac{\partial K}{\partial \psi}(\dot{\tau} + l\dot{\psi} + m\dot{\vartheta} + n\dot{\varphi}),^2 \end{aligned} \tag{331}$$

and the last member is identified as the negative of the corresponding quantity in equation (330). Its appearance in the final forms is intimately related to a diversion of energy that persists, though the action of (Q) is veiled otherwise. Utilizing all these detailed relations justifies the equality, where the notation for the last term in the first member indicates the condition observed, and for (dl/dt) we have inserted the value

$$\frac{dl}{dt} = \frac{\partial l}{\partial \psi}\dot{\psi} + \frac{\partial l}{\partial \vartheta}\dot{\vartheta} + \frac{\partial l}{\partial \varphi}\dot{\varphi},$$

$$\left.\begin{aligned}
&\frac{d}{dt}\left(\frac{\partial E_{(0)}}{\partial \dot{\psi}}\right) - \frac{\partial E_{(0)}}{\partial \psi} + Q\left[\left(\frac{\partial l}{\partial \varphi} - \frac{\partial n}{\partial \psi}\right)\dot{\varphi}\right. \\
&\quad \left. + \left(\frac{\partial l}{\partial \vartheta} - \frac{\partial m}{\partial \psi}\right)\dot{\vartheta}\right] + \left[\frac{\partial E_{(Q)}}{\partial \psi}\right]_Q = \frac{d}{dt}\left(\frac{\partial E}{\partial \dot{\psi}}\right) - \frac{\partial E}{\partial \psi} \\
&\quad + \left[\frac{\partial E_{(Q)}}{\partial \psi}\right]_Q = \frac{d}{dt}\left[\frac{\partial}{\partial \dot{\psi}}(E_{(0)} + E_{(Q)})\right] \\
&\qquad\qquad - \frac{\partial}{\partial \psi}(E_{(0)} + E_{(Q)}) = F_{(\psi)}.
\end{aligned}\right\} \quad (332a)$$

To which the companions added after cyclic interchanges are

$$\left.\begin{aligned}
&\frac{d}{dt}\left(\frac{\partial E_{(0)}}{\partial \dot{\vartheta}}\right) - \frac{\partial E_{(0)}}{\partial \vartheta} + Q\left[\left(\frac{\partial m}{\partial \psi} - \frac{\partial l}{\partial \vartheta}\right)\dot{\psi}\right. \\
&\quad \left. + \left(\frac{\partial m}{\partial \varphi} - \frac{\partial n}{\partial \vartheta}\right)\dot{\varphi}\right] + \left[\frac{\partial E_{(Q)}}{\partial \vartheta}\right]_Q = \frac{d}{dt}\left(\frac{\partial E}{\partial \dot{\vartheta}}\right) \\
&\quad - \frac{\partial E}{\partial \vartheta} + \left[\frac{\partial E_{(Q)}}{\partial \vartheta}\right]_Q = \frac{d}{dt}\left[\frac{\partial}{\partial \dot{\vartheta}}(E_{(0)} + E_{(Q)})\right] \\
&\qquad\qquad - \frac{\partial}{\partial \vartheta}(E_{(0)} + E_{(Q)}) = F_{(\vartheta)}; \\
&\frac{d}{dt}\left(\frac{\partial E_{(0)}}{\partial \dot{\varphi}}\right) - \frac{\partial E_{(0)}}{\partial \varphi} + Q\left[\left(\frac{\partial n}{\partial \psi} - \frac{\partial l}{\partial \varphi}\right)\dot{\psi}\right. \\
&\quad \left. + \left(\frac{\partial n}{\partial \vartheta} - \frac{\partial m}{\partial \varphi}\right)\dot{\vartheta}\right] + \left[\frac{\partial E_{(Q)}}{\partial \varphi}\right]_Q = \frac{d}{dt}\left(\frac{\partial E}{\partial \dot{\varphi}}\right) \\
&\quad - \frac{\partial E}{\partial \varphi} + \left[\frac{\partial E_{(Q)}}{\partial \varphi}\right]_Q = \frac{d}{dt}\left[\frac{\partial}{\partial \dot{\varphi}}(E_{(0)} + E_{(Q)})\right] \\
&\qquad\qquad - \frac{\partial}{\partial \varphi}(E_{(0)} + E_{(Q)}) = F_{(\phi)}.
\end{aligned}\right\} \quad (332b)$$

146. It is plain from these forms how the actual values of the last members but one for the energy changes in the system may be preserved and an account of them be given under various other interpretations that are in a sense fictitious. Or they are put in a fashion that uses knowledge up to its borders, with safe non-committal beyond them. What is here exemplified for one coördinate ignored, can be extended of course to many by a similar procedure. And when acceptance of reduction factors has widened the range outside that covered by the geometrical direction cosines, intricacies of energy connections are made resolvable in many general ways.

It may happen that some contributions to the total group of forces acting on a system are comprised under a potential energy function; and it is in the nature of those relations that such forces are independent of velocities. If therefore there is any gain in doing so, the active forces may be held asunder in two groups, one containing all the forces derivable from any potential energy functions (Φ). Then in any coördinate (κ) the new model of Lagrange's equations is only formally varied when it is written

$$F'_{(\kappa)} = \frac{d}{dt}\left(\frac{\partial}{\partial \dot{\kappa}}(E - \Phi)\right) - \frac{\partial}{\partial \kappa}(E - \Phi), \qquad (333)$$

since (Φ) is inoperative in the first term, and in the second it only transposes one group of the forces. But this type offers the significant feature that a course of events to which the first member can be the key, is exhibited as depending upon the momentary outstanding difference between two quantities measurable as energy. And with the door opened as usual to seemingly vital analogies among energy forms, much is being done in these days to increase the command of dynamical statement for the most inclusive rules or principles deciphered among physical sequences of transformed energy. It did not seem, therefore, that the objects of the chapter on the side of stimulating suggestion would be attained unless we were brought to this gateway into a larger field. But then too we must be content with that much of accomplishment, leaving the other forms of Lagrange's equations, beside this second one as they are usually counted, to the systematic continuations of which there is no lack. The exploitation of the concept called *kinetic potential*, whose roots can be traced in the difference (E − Φ), and its alternative origin as a deduction from Hamilton's principle of *stationary action*, are the groundwork of much modern dynamical thought.[1]

[1] See Note 32.

Notes to Chapters I–IV.

Note 1 (page 2). To be aware of are an initial trend through the drift impressed by the nature of the material, as well as an active later movement with its propaganda. Regarding the first of these headings it is discussible whether the opinion alluded to in section 3 is fully representative of Newton's own standpoint, or whether that tendency to one-sided development was due to adherents whose acceptance of ideas was narrower than the scheme of his proposal. So much can be done by way of expanding or contracting the thought lying behind a condensed formulation in Latin that we tread on insecure ground in attempting a decision. Safest it seems to allow in Newton's plan at least potential provision inclusive of all that two succeeding centuries could reasonably urge on this score. Adding perhaps, what expert judges would have us not overlook, that a comprehensive power-equation is laid down in the scholium to the third law. Read in English thus: "If the Activity of an agent be measured by its amount and its velocity conjointly; and if, similarly, the Counter-activity of the resistance be measured by the velocities of its several parts and their several amounts conjointly, whether these arise from friction, cohesion, weight or acceleration;—Activity and Counter-activity, in all combinations of machines, will be equal and opposite" (Thomson and Tait, Natural Philosophy (1879), Part I, page 247). The genius of Heaviside for directest dynamical thinking approves this scholium as capable of covering the fluxes and transformations of energy that more recent dynamics introduces (Electromagnetic Theory, III, pages 178–80).

In the movement toward basing the derivation of other concepts upon energy, Tait put forward an early denial of primary quality to force in a lecture before the British Association (1876). The habits of thought in these respects, however, are interwoven with a widespread campaign extending over the main issues of epistemology (Erkenntnistheorie) that enlivened the period 1895–1905, some of whose other aspects are touched upon subsequently (see notes 4 and 5). The party there whose watchword was "Phenomenology" made common cause with energetics as a properly neutral mode of statement, in opposition to theoretical physics—or more justly to overweight in speculation. These matters of broad sweep are only to be hinted here; they are fully in evidence throughout the journals of that date. Yet we may admit mention of two books, one showing how energetics counterpoises and supplements other aspects of dynamics, and the second exhibiting by contrast exaggerations into which zealous advocates were led. The titles are: Helm, die Energetik (1898); Ostwald, die Naturphilosophie (1902).

Note 2 (page 4). The spirit of this paragraph finds confirmation in recent judicial utterances, as regards both appreciation of the new movement and prudent reserve in passing judgment. Consult Silberstein, The Theory of Relativity, for a lucid account of the Lorentz-Einstein method that estimates its gains with candor and acumen. The workable value in the opened vein of possibilities will be extracted progressively, as its logic is brought to bear upon questions involving previous sequences and their origins. Poincaré expresses this plainly in his summing up: "Aujourd'hui certains physiciens veulent adopter une convention nouvelle . . . plus commode, voilà tout. . . . Ceux qui ne sont pas de cet avis peuvent légitimement conserver l'ancienne. . . . Je crois, entre nous, que c'est ce qu'ils feront encore longtemps" (Dernières Pensées, page 54). Clarification and settlement here seem delayed by an observable tendency to expound the central ideas of relativity in an entanglement with much irrelevant mathematics that is describable also as transcendental. This blurs essentials and will obstruct the final rating of the novel features among the resources of physics. It is foreign to such alliance, and hence perhaps one influence toward dissolving it, that the modified handling of simultaneousness traces its lineage so directly to experimental evidence, and the effort to state its results with unforced symmetry. Yet on that side, too, there might arise need of corrective, if perchance the conclusion were entertained seriously, that any newly assumed attitude releases us from that bondage to idealized concepts and simplifying approximations which sections 12 and 13 indicate. We should be compelled to reject every inference that some system invented to replace Newtonian dynamics can be other than differently conceptual and approximate. What alternative concepts to employ will always remain as a choice determined on practical grounds. It would be breaking with the canons of sound scientific doctrine to displace one series of working ideas by another whose improved adaptation to universal service is at best to be classed among open questions. Though symmetry in equations is desirable, it is not to be secured at all costs. In order to turn the balance conclusively, insight must first be attained that goes far enough in excluding illusion from the corresponding dynamics. The characteristic formulas of relativity draw their suggestion from groups of phenomena that spread over limited area as compared with the explored range of physics. Their analysis beyond the kinematical stage, moreover, is too obscure and intricate as yet to afford mandatory reasons, or even trustworthy guidance, for much reshaping of our fundamental equations. See note 11 below, in continuance of this thought.

Note 3 (page 6). The reference is to Maxwell's Treatise on Electricity and Magnetism, II, Chapters V and VI of Part IV. He records (1873) the stimulus received from the Natural Philosophy by Thomson and Tait (1867), and from the revival of dynamical advance inspired by "that stiff but thoroughgoing work" (Heaviside). It continues to offer an unexhausted mine to a later generation. In its second edition (1879) the present topic by added material and recasting points rather plainly toward mutual reaction between Maxwell and its authors. It is true that their expanded treatment does not explicitly occupy his larger field, though their gyroscopic illustrations run easily, as can be seen, into a generalized scheme of cyclic systems. In that direction Ebert, with Chapters XX–XXII of his Magnetische Kraftfelder (1897), has made a junction by elaborating into dilution the results of Hertz and Helmholtz. Others like Gray prolong directly the line of Maxwell's initiative (Absolute Measurements in Electricity and Magnetism, II, Part I, Chapter IV (1893)).

It is not premature to remark, in anticipation of notes 30 and 31, and with bearing upon the current presentations of Lagrange's equations, how guardedly the vectorial connections of their original scope are relaxing. We may suppose that the freedom to cut loose in this respect has been for a time masked by the cartesian (XYZ) forms, whose effective reduction to quasi-scalar expressions has had an influence elsewhere, as pointed out in section 91, toward indifference about such distinctions that fails to regard them as vital.

Note 4 (page 9). What is appropriate here in preparing for intelligent command of stock resources must not go far beyond claiming for these inquiries a continued relation to the organic structure of dynamics of which their perennial life is one convincing proof. Some study of their literature cannot be dispensed with, from which differently shaded opinions will be drawn, to be sure, that will yet unite in agreement on the final importance of the answers. To recommend this as one region for deliberate thinking is the purpose at this place, leaving opinions to shape themselves individually. The concession how fully routine belonging to execution can go its way unhampered by deeper questions should be permitted to repeat itself without undermining finally the need incumbent upon us to discuss them. Section 16 alludes to some temporary grounds for unconcern, others are supplied by the sufficiency of a fixed earth's surface for staging so many investigations of physics, and in various directions a fortunate postponement is tolerated. But testimony is broadcast how steadfastly some settlement is nevertheless held in view, for the experimental bearings of it even, when freed from all metaphysical residue. For exemplifying reference take Larmor's comment (Aether and Matter, page 273) and Helm's pertinent remark (Energetik, page 216).

There were several leaders in the public sifting of these theories. Prominent among them Mach, who has gone on record in his Science of Mechanics, Chapter II, and elsewhere.* The possibility of the so-called Newtonian transformation having been put on a secure basis, that headed unconstrainedly toward using an origin at the center of mass of the solar system and directions determined by the stars for a natural reference-frame. Especially for what are rightfully classed as internal energies of the system this would be capable of high precision in presenting through accelerations relative to it, for the bodies with which we deal, the physical forces active among them or upon them (see section 52, and note 17). It is a live question of the passing time whether that habit of mind had better be upset, or can be superseded with definite net gain.

Note 5 (page 17). The assertion is hardly contestable, that quantitative physics deals with an idealized and simplified skeleton built of concepts, so soon as its content exceeds the rules that are empirical by intention and form. The supports found for outstanding argument are then two: first, uncom-

* This is the briefer title of the English translation, the original title being "Die Mechanik in ihrer Entwickelung historisch-kritisch dargestellt."

promising denial that the goal can be aught else than empirical rules, ingenuity being restricted to embodying best in them the ascertained data; or secondly, in questioning doubt how the boundary-line runs among special cases. Troubles of the latter origin involve no radical divergencies, since they are everywhere inherent in such a separation of two classes, both being acknowledged to exist. Positions like the first mentioned would be a fetter upon growth through their exclusive blindness to patent and historic facts, were not a saving clause inserted in extremist tenets by human readiness to lapse into inconsistency for good cause. To illustrate how the main contention spoken of would cramp effort, we find place for a quotation, which however is content to set two standards in opposition: "Die Fourier'sche Theorie der Wärmeleitung kann als eine Mustertheorie bezeichnet werden. Dieselbe . . . gründet sich auf eine beobachtbare Tatsache nach welcher die Ausgleichungsgeschwindigkeiten [kleiner] Temperaturdifferenzen diesen Differenzen selbst proportional sind. Eine solche Tatsache kann zwar durch feinere Beobachtungen genauer festgestellt werden, sie kann aber mit andern Tatsachen nicht in Widerspruch treten. . . . Während eine Hypothese wie jene der kinetischen Gastheorie . . . jeden Augenblick des Widerspruchs gewärtig sein muss" (Mach, Prinzipien der Wärme, page 115). We know that the goal here implied for theory is only the starting-post for it in the doctrine of another school of thinking; but must abstain from even outlining the argument.

The important concern for dynamics here turns plainly upon the question of aligning it in method with the rest of mathematical physics, or of excepting it from partnership in a search for confessedly empirical rules. In point of fact, this one undeniably fruitful wielding of idealized conditions has been a bulwark of defense for universal procedure. No interested student can afford to neglect Poincaré's pronounced judgment in this field, to be found especially under the four book-titles: La science et l'hypothèse; La valeur de la Science; Science et Méthode; Dernières Pensées. The first three are most compactly accessible in one volume of English translation headed The Foundations of Science (1913); the fourth not included in that collection is of recent date (1913) and presents much that is of value. Far from putting these matters aside as completed, latest developments have renewed and intensified their lively discussion. As representative in one direction we name the work of Robb: A Theory of Time and Space (1914); and on another line a paper by N. Campbell (1910), The Principles of Dynamics (Philosophical Magazine, XIX, page 168). These will sufficiently lay out a track for further pursuit, in connection with notes 1, 4 and 6.

Note 6 (page 24). There is much more here than the kinematical colorlessness that precedes the introduction of dynamical elements. Attention is being directed to that stage of inclusive preparedness in the fundamental equations that is one permanent attribute of "Analytic mechanics," in so far as its forms of statement are made equally ready to contain various specialized data. Workers in the subject really avail themselves of this privilege to delay in particularizing. Lorentz for example does not attempt to settle in advance which reference-frames meet the conditions attached to the primary relations for the electromagnetic field. He lays the decision aside temporarily with the passing remark that the equations remain valid so long as they accord with the value ($c = 3 \times 10^{10}$ cm./sec.) for light-speed in free space. So a top's local behavior relatively to the earth's surface follows equations of motion in common with the gyroscopic compass up to a certain divergence-point, though the former ignores the earth's rotation, and the latter may be said to reveal it. In a group of parallel cases the differences center upon replacing gravitation by weight; which illustrates how essentially the standards of desirable or attainable precision enter into adapting broader analytic expressions.

Note 7 (page 26). A number of points touching the fuller incorporation of vectors into physical purposes must become more definite presently, as the novelty of their use subsides. Conventions that have been transferred from mathematical definitions, or that have been added tacitly, will be opened to needed revisions first by being made explicit. The text will be found to adopt this feature of sound policy at several places, none of which should be slurred. Care to *delimit equivalences legitimately* in relation to physical conclusions is one leading idea as regards substitutions that approves itself to be a needed refinement upon the looser term *equality*. For accelerating the center of mass of a system forces have the quality of free vectors, because their position is without effect upon equivalence in this respect. Yet when we discuss motion relative to the center of mass, forces fall away from that equivalence, being then dependent upon position for their effect, and consistently they cease to be free vectors. Such instances compel us to qualify classifications and permissible substitutions.

Similar deliberateness in borrowing from mathematics is encouraged in section 68, with its suggested distinction between triangle and parallelogram as graphs of a vector sum; and in section 74, where an element of parallel shift enters to round out the variableness of a vector quantity.

The idea of vector-angle used in equation (2) has not yet found its way into textbooks. Its introduction is an almost self-evident detail of any systematic vector algebra, to supply the missing member of the series in which angular velocity and acceleration were long since recognized. How that proves helpful is elaborated in section 92 and its sequel. The simple step of completing with natural orienting unit-vectors the established ratio (ds/r) for magnitude of angle seems to be announced first in the Physical Review (N. S.), I, page 56 (1913). In section 46 the text opens from this side a new meaning for the rotation-vector that fits usefully in several ways, though it is, of course, nothing but that second interpretation possible for every vector product which happens to have been overlooked here. We must ascribe the oversight to a continuance of the earlier exclusive habit of using only the projection of ($\mathbf{r}$) that is perpendicular to ($\boldsymbol{\omega}$), and not the corresponding projection of the latter vector.

Notice how the rotation-vector can be given another rôle if we rewrite equation (44) in the form

$$\mathbf{v} = -(\mathbf{r} \times \boldsymbol{\omega}),$$

reading the second member as the negative moment of ($\boldsymbol{\omega}$) distributed locally at each (dm). This has important connections with the uses of vector potential, and the association of the curl operator with the latter.

Note 8 (page 31). Later research has come to the aid of mathematical demands or convenience on this side, by detecting real transitions with however sharp gradient behind most first assumptions of discontinuous break. In proportion as facts of that character gather they soften the impression of artifice in making phenomena amenable to treatment by allowing for quick gradations, and incline modern physics away from recognizing discontinuous change except upon compulsion. See Lorentz, The Theory of Electrons (1909), page 11. This accounts probably for some psychology alongside the mathematical needs mentioned in section 26, of which we might admit an admixture in the satisfaction, when identity preserved or at least quantity conserved is attributable anywhere without too strained devices. Poincaré's shrewd remark is to this effect: "Physicists can be relied upon to find something else whose total remains invariant, should energy leave them in the lurch." And is there not some shade of disappointment in conceding our failure to trace individual elements of energy by Poynting's theorem, as well as the paths of flux? Compare Lorentz, The Theory of Electrons, page 25; Heaviside, Electromagnetic Theory, I, page 75 (1893).

Note 9 (page 33). To follow lines that are accommodated to some directive idea of constancy gives in many ways a natural order. About this we should acknowledge though, how inevitably our assigning conceptually common or constant values takes its suggestion from what are means or averages in their experimental basis. Neither must the truth be forgotten with which section 69 closes. The enlargement in application through free use of mass-averages, time-means, and the like can be instanced for the immediate connection from sections 20, 21 and 31. But it confronts us without any special search everywhere in physics, when we remember that the point at which values are admitted to be "local" is in practice solely a matter of scale; they are finally representative of mean values to a certain order of precision (compare section 42). Less familiar but perhaps just as significant is that reading of the curl and the divergence locally in a vector field which sees in them the specification of an artificial symmetry which rests upon mean values, and replaces legitimately for certain ends the actual field-distribution. See the Physical Review, XXXIV, page 359 (1912); Boussinesq, Note sur le potentiel sphérique, pages 319–329, in his Application des Potentiels à l'étude de l'Équilibre et du Mouvement des Solides élastiques (1885).

Note 10 (page 36). Every such element that is force-moment presents a local resultant, similar to those met in section 19 through being normal to the individual plane of its factors. As vector products these local resultants are all open to the same sort of double reading as is brought up for the rotation-vector in note 7 and completed in note 16. The process of mass-summation for a system then continues associated with the resultant elements (d**H**) or (d**M**), combining each group as a vector sum to a total resultant of determinate orientation and tensor. The fraction of this last resultant effective or available in relation to any particular axis of unit-vector ($\mathbf{a}_1$) is ascertainable by one final projection, representable respectively by

$$\mathbf{H}_{(\mathbf{a})} = \mathbf{a}_1(\mathbf{H}\cdot\mathbf{a}_1); \quad \mathbf{M}_{(\mathbf{a})} = \mathbf{a}_1(\mathbf{M}\cdot\mathbf{a}_1).$$

The departure from the cartesian scheme consists especially in reserving projection for the closing operation, to be executed only when the demand for it enters. There is the common inversion of order between mass-summation and projection, on passing over to vector algebra.

Note 11 (page 43). There is a considerable region opened to plain sailing among developments like those of sections 32–35, whenever the observed material justifies our major premiss that inertia occurs as a variable quantity. But whatever general bearings may be obtained thus, we do not of necessity touch the source of the inferred variableness, and much less do we reach a halting-place about it in default of supplementary evidence. The emphasis of the text is focussed upon the truth of this remark which is of wide application, the electronic case being included among others. Consequently there is a warning implied to avoid a pitfall: ascribing prematurely the appearing variableness to one type of source among several of which experience has made us aware, and thereby affecting the conclusions with fallacy. The conscious fictions that cluster round the idea of effective mass should make us wary of deceptive illusions there whose enigma has not been resolved. The capacity of an unincluded (or undetected) force to compel indirect recognition of itself in the inertia-coefficient is well-known. And a long line of suggestive connections with processes of continuously repeated impact have their root in an old problem. This is the transmission of elastic deformation through a bar struck at one end (see Clebsch, Théorie d'Élasticité des corps solides, translated by St. Venant, page 480a, Note finale du § 60). A possible modification of that treatment for impact has been set forth repeatedly, in the attempt to cover wider conditions of converting and storing energy within a system, under some form of structure or arrangement. Heaviside especially has achieved instructive results under that heading. The cogency of the logic in transferring demonstrated consequences of this nature to electrons hinges on the query in how far the convective energy of electromagnetic inertia is adequately analogous to the kinetic energy of (ponderable) mass. At this date it would plainly beg the larger question to assert unreservedly that both these forms of energy are literally the same.

Note 12 (page 44). In the closing chapter of his Kritische Geschichte der allgemeinen Prinzipien der Mechanik (1877) Dühring urges the sound advice not to stop short of first-hand contact with the notable contributions that mark epochs of advance. The case of d'Alembert's discovery enforces the wisdom of that counsel, because a tradition echoing an imperfect apprehension of the principle has leaned toward perverting the gist of it from the meaning that the leaders in dynamics state clearly, whose essential thought sections 38–41 aim to restore. Compare them with the analysis of the principle in Mach's Science of Mechanics and in Helm's Energetik. One source of confusion can be located in the transposition that yields the forms of equation (38). This point is alluded to at the end of section 41; and the idea is expanded with elementary detail in Science, XXVIII, page 154. Some obstacles to ready understanding are due no doubt to a certain crabbed brevity of the nascent formulation in d'Alembert's Traité de Dynamique (1758), found in Chapter I of Part II. A German translation of this classic is provided among Ostwald's Klassiker der exakten Wissenschaften (Number 106).

Note 13 (page 51). The influence of the energetic view pervades the handling of energy flux and of the accompanying forces or stresses. The transfer-forces of the text appear for example in Helm's exposition (Energetik, page 233 and *passim*). The habit of thinking in these terms is cultivated by greater familiarity with storage of energy in media, which has added the vigor of a physically conceived process to the formal nature of potential energy in the earliest instance of gravitation, where the mechanism remains completely obscure (see section 3). It is growing increasingly evident how the outcome of explorations among energies intrinsic and external is capable of reduction in parallel fashion, exhibiting the conditioned modes of revealing their presence and the measured extent of their availability. The lessons about cautious inference of which some scant mention is made in the text are perhaps nowhere more impressive among the inductions of physics, when once the safety of non-committal attitude must be abandoned in active search for a determinate process. We remember the remark that "An infinite number of mechanical explanations are possible" (Poincaré), especially since we deal primarily with finite or statistical resultants; and even plausible schemes are numerous enough to leave a broad margin for indecision. See Lorentz, Theory of Electrons, pages 30–32.

Poynting's original paper should not be left unread (Philosophical Transactions (1884), Part II, page 343); nor touch be lost with Heaviside's stimulating directness (e. g., Electromagnetic Theory, I, pages 72–78). A sensible summary incorporating links with relativity is furnished by Mattioli; Nuovo Cimento (series 6), IX, pages 255, 263 (1915).

Note 14, page 53. Geometrical conditions are always a needful auxiliary in expressing constraints for the reason named in the text. The use of indeterminate multipliers would carry unreduced geometrical forms into the equations of motion, giving what might be called quasi-forces. Lagrange himself offers that analysis of their significance in his Mécanique Analytique, I, pages 69–73 (Bertrand's edition (1853)). Later practice runs more nearly in the line of separating these supplementary relations from the purely dynamical truths, and using the former admittedly as mathematical aid in eliminations looking to ends like integrations. Thomson and Tait held it part of their service to have brought together the fully dynamical treatment of constrained and of free systems (Natural Philosophy, Part I, pages 271, 302).

Note 15 (page 54). The point now reached offers occasion to add explicit reference to Routh's encyclopedic work in two parts: Elementary Rigid Dynamics, Advanced Rigid Dynamics; as a storehouse to which we shall long resort for authoritative presentation of characteristic material in this field. The design of our text has acknowledged as one main object to foster the study of masters such as Kelvin, Routh and a few others in dynamics. To this end we are building a less steep approach to the level upon which their progress moves. It cannot be said to stand in prospect that these writers will become antiquated; but need will arise from time to time for seeing the older systematic grouping in an altered perspective, in order to renew connections or symmetry that temporary stress upon some lines of growth may have disturbed.

Note 16 (page 57). Preparation has been made by anticipation in the connection of notes 7 and 10 to accept this meaning and office for the rotation-vector which are an enlargement upon the usual current statement about it. That aspect is adapted to set in higher relief its comprehensive and yet particular relation to those individual radius-vectors upon which vector algebra turns attention. There is some advantage gained, too, by approaching the special rigid connection on the line that starts with the complete freedom in equation (2), and sees the vector (ω) of common application to all radius-vectors to be an outgrowth of that rigidity.

Note 17 (page 64). It is important to keep track of successive restrictions that enter to affect the range of conclusions. Here we must not overlook that the added condition of rigidity influences only a *general reduction in form* for certain parts of (E, **H**, P, **M**) that are seen to occur already in equations (10, 12, 54, 55) as written for any non-rigid system of constant mass. In brief, the notion of a constituent translation with the center of mass applies to all such systems; and so does the independent treatment of that translation and of the motion relative to the center of mass, as spoken of in section 52. That point is elaborated for elementary purposes in my Principles of Mechanics, Part I, pages 91–101. Including now equations (19, 20) it is made fully evident how no new situation is introduced when we ascribe rigidity to the body, except in the entrance of rotation. While absorbing the *residual* (E, **H**, P, **M**), this type of motion also gives concise expression to their values, in every one of which, it will be noticed, either ($\boldsymbol{\omega}$) or ($\dot{\boldsymbol{\omega}}$) appears, marking the relation of both to the body as a whole.

Note 18 (page 70). The frequent necessity of a dynamically active couple for an adjusted control securing kinematical constancy in the vector ($\boldsymbol{\omega}$) is now an everyday lesson learned from the *directive couple* of rotation about a fixed axis. The possible divergence of ($\boldsymbol{\omega}$) and (**H**) furnishes the simple key which cuts off vector constancy of both together; with habitual demand then prevailing for some (**M**) associated with every change in (**H**). But there has been an astonishing record of tenacious refusal to distinguish between such conditions of active control and the conditions of equilibrium, here and in the companion instance of radial control requisite for continuance of circular motion. The surviving power of instinctive prepossessions has perpetuated in unexpected quarters the ancient unclearness lurking behind "centrifugal force and couple"; and this threatens to endure under the full illumination of the vector view. The root of many like confusions is traceable to a failure really to grasp the facts in the first of equations (38), with unfaltering discrimination between impressed and effective forces. That equation does not describe an *actual equilibrium;* neither does the result of any transposition which yields an equation like the second form of (77). Yet compare the presentation by authorities: Klein and Sommerfeld, Theorie des Kreisels (1897), pages 141, 166, 175, 182; though no criticism applies anywhere to their mathematical correctness.

Note 19 (page 82). This labored insistence upon the dual aspects of all coincidences is indeed designed to remove an ambiguity in symbolism whose currency has grown out of imperfect attention to them. There is usually reward for watchfulness on those points. But the allowableness of such detail in the text rests more upon its initiative for developing the idea of shift in section 79. Notice, as we proceed, how often the unit-vectors and the tensors of vector quantities offer themselves naturally as independent variable elements, and afford a ground for partial differentiations of a type peculiar to vector algebra.

Note 20 (page 88). Of course forces are "bound to superposition" only by the same tie of definition or specification that holds velocity and acceleration also, and that is broken when we abandon the parallelogram graph. But it is remarkable how regularly in physics that mutual independence among energies (and among forces that change them) is experimentally supported, of which superposition and linear relation are mathematical expression. Still it is reasonable to grant that not all definitions devised for physical quantity have escaped a bias from this side which will need to be allowed for or rectified. Yet the high price paid for relinquishing that simplest rule warrants the change of base only on clearest showing of the balance-sheet.

By referring to "physical status" the text means to encourage that scrutiny for terms of algebraic origin whose favorable and unfavorable outcome in particular connections it cites in several places. To be sure, candor and detachment are called for continually in reaching judgment through the arguments by convergent plausibility upon which closing of the doubtful issues here depends (see sections 6 and 7).

Note 21 (page 93). The superficial features of what is here named shift are detectable generally in previous accounts of coördinate systems; and Hayward is often credited with a comprehensive survey of the subject in a paper: On a direct method of estimating velocities with respect to axes movable in space (Cambridge Philosophical Transactions (1864), X, page 1*). Anticipations of the controlling purpose in shift might be expected confidently, since its ramifications are now recognizable through all that coördinate machinery of early devising without which commonest operations of algebra would have been blocked. But the circumstance seems exceptional that completed analysis of its working has been postponed. The proposition presented by equation (137) does not occur in the first editions of Routh, and he never gives to it deserved prominence. Abraham's statement of it is of course formally right, yet he describes our (X′Y′Z′) questionably as a "Rotierendes Bezugssystem" (Theorie der Elektrizität (1904), I, page 34). The relations of coincidence that make equation (124) important Routh disposes of in one obscurely placed line: "As if the axes were fixed in space" (Elementary Rigid Dynamics (1905), page 213). Equally casual is Abraham (p. 115): "Die Umrechnung [auf ein bewegtes Bezugssystem] geschieht genau so, als ob das bewegte System in seiner augenblicklichen Lage ruhte." This comparative blank left place for that more systematic or conscious display which vector algebra favors of the really operative methods. Its

* This is the date of publication. The paper itself was dated and read (1856).

partial novelty has set its measure at a length in the text that may well be curtailed when their leading thought has once been laid down.

Note 22 (page 98). Some authors cover the point by a distinction between explicit and implicit functions of time. Or again the changing relation fairly equivalent to our shift of (**i′j′k′**) among (**ijk**) is made to introduce a partial time-derivative (Thomson and Tait, Natural Philosophy, Part I, page 303). It cannot escape notice what direct gain in clearness the regular acceptance in our algebra of time-derivatives for unit-vectors yields. The due adjustment of pace for shift, especially in order to simplify dynamical problems in astronomy, has called forth important discussion touching the double entry of time, while methods of treating perturbations were becoming fully established; and this engaged the attention of men like Donkin, Jacobi, Hansen. There is a sequel in that region to sections 107–112; see, for instance, Cayley, Progress in Theoretical Dynamics, British Association Report (1857).

Note 23 (page 109). The type to be remarked in equations (154) as leading to generalizations of them is the functional relation between each of ($\dot{x}'$, $\dot{y}'$, $\dot{z}'$) and all of both ($\dot{x}$, $\dot{y}$, $\dot{z}$) and (x, y, z). The same combinations show reciprocally when equations (150) are differentiated, and they affect characteristically the expressions derived for kinetic energy. In equations like (155) the first equality of partial derivatives brings out the extent to which building up is occurring in the instantaneous lines of (x′, y′, z′); and the second such equality connects the remainder of the increment visibly with changes of slope that are proceeding. It becomes then a simple matter to forecast how these constituents will reproduce the result given through a vector derivative.

Note 24 (page 118). One main objective being to specify configurations in the standard frame, it is indispensable in the plan that some unbroken link with the latter should be maintained. The permanent orientation in (Z) of the angle-vector (ψ) serves that purpose, every displacement $(d\psi)$ being immediately relative to (XYZ). By the terms of section 93 displacements in (ϑ) have this one step interposed between direct junction with (XYZ); and finally displacements in (ϕ) are two removes from that immediate relation. Taking other comment from the text, it is made apparent how adequately all this parallels the conception of displacements parallel to (X, Y, Z) as successive, independent, and cumulatively relative. There too, whichever the second and third displacements are, according to the order selected, each must accept a determined initial state due to the displacements that have preceded it. The residual difference is inherent in the mutually supplementary qualities of linear and angular displacements. Other parallel features with longer-established vector schemes will repay attention; for example the sentence just preceding equation (174) does not mark an exceptional condition. It is of interest, too, to dwell upon the fact implied on page 120, that (ψ, ϑ, ϕ) give us the model of a coördinate-set with a changing obliquity among its unit-vectors. It is obviously unessential, except for convenience, that $(\mathbf{i}'\mathbf{j}'\mathbf{k}')$ should be orthogonal or retain any constant relative obliquity. Some proposals have been made to include the more general relation of direction for sets of unit-vectors; and the necessary modification of section 45 would be no more than simple routine.

Note 25 (page 125). Needless to say, the revised conclusion reached through equation (186) renounces any attempt to make complete derivatives out of what are actually partials; but it succeeds in assigning their proper quality to derivatives, for all such combinations involving vectors, under a general rule stated at the close of section 100. The root of the matter goes back to equation (124); and the establishment of angle among vectors places it in a category with them in this respect also. In what form the omission of that element raises the difficulty may be gathered from Klein and Sommerfeld, Theorie des Kreisels, page 46. The truth is that a similar non-integrability of tensor accompanies every plan of shift, except those in which a special condition is satisfied that includes them among what may be classed with *envelope solutions* (see section 116).

Note 26 (page 137). The text bears frequent testimony consistently to a high appreciation for the genius and inspiration of the earlier workers who built dynamics, among whom we may name Coriolis. Yet we should respect our obligation also to carry forward or to rectify the first suggestions; being taught to expect advances in our reading attached to results especially, whose mathematical accuracy has never been questioned. It is that hint of possible improvement which the text here submits, affirming the lesson of cultivating perception of physical meanings upon which best modern thought concentrates, and which is illustrated by sections 35, 57 and 104; all to be taken in the light of repeated comment upon those clouding transfers between the two members of equation (37) which are still too prevalent.

Note 27 (page 141). Hansen, Sächsische Gesellschaft der Wissenschaften, Mathematisch-physikalische Klasse, III, pages 67–71. This original statement retains value, partly still through the material it discusses, and again through the moral it conveys that vector methods have made these problems more manageable. The reaction of Jacobi in some letters to Hansen (Crelle, Journal für reine und angewandte Mathematik, XLII, (1851)) shows instructively the struggle toward clear and firmly grasped thought proceeding, with strictest scrutiny of detail in the new proposal. In the paper referred to above, Hansen's double use of time is worked out (compare note 22), that remains current among astronomers.

Note 28 (page 155). We do not measure rightly the inheritance of rigid dynamics from Euler's labors without conscious effort to reconstruct the void that they filled once for all. Unless his inventive intuitions had here been favored by a happy chance, he could hardly have moulded from the first heat so many of the forms that seem destined to hold permanent place. We can imagine that his inspiration caught early glimpses of the relation that equations (72) and (258) now convey; but Euler may have been content to seize the validity of equation (257) without proving it, as Fourier did in like case. Certain it is that the point involved in that equivalence seemed troublesome enough to be made the object of various special proofs, before our general equation (137) had been attained (see Routh, Elementary Rigid Dynamics (1882), page 212). For the historic date, the memoir presented to the Berlin Academy is quoted (1758). But a satisfactory survey of Euler's contributions on the topic is best obtained through his collected works. Easier access perhaps is had in the German translation (Wolfers, 1853); in the volumes 3–4 entitled Theorie der Bewegung the "Centrifugal couple" appears at page 323, and our main interest would probably concentrate on pages 207–443.

Note 29 (page 169). Klein and Sommerfeld, Über die Theorie des Kreisels (1897–1910), is one instance, quoting our Preface, how special treatises of unquestioned excellence make superfluous an attempt to replace them. This work, and Routh's version in the Advanced Rigid Dynamics (edition of 1905), Chapter V, with Thomson and Tait's discussions *passim* in Natural Philosophy, Part I, supply for gyroscopic problems the indispensable material, exhaustive of more than their general aspects. The aim of the text here is strictly confined to lending its announced special emphasis to two items. One is shown to be of ramifying importance as a singular value round which deviations from it may be organized; the other is uniquely characteristic, and it proves amenable to this analysis most simply, in comparison with other methods. Compare in verification Theorie des Kreisels, pages 247, 316, on strong and weak tops.

Note 30 (page 180). A fuller command of generalized coordinates and forces as an effective working method can be inferred from evidence on two sides: first, more unequivocal recognition is accorded to their finally scalar type; and secondly, the primary demonstration of relations shows increasingly directer insight. Dispose of the latter point by collating Lagrange's proof (Mécanique Analytique, I; Dynamique, Section IV); Thomson and Tait, whose change between (1867) and (1879) is instructive; and Heaviside, Electromagnetic Theory, III, page 178. The last-named is a climax of condensation, and thereby somewhat unfitted for the text; but it will be quoted below for a double reason. The quantitative emancipation of Lagrange's equations may be traced gradually, if we like, beginning with equations such as (150, 151), where the (l, m, n) coefficients are particular reduction factors conditioned as in equation (152). Next advance to the more liberal possibilities of linear vector functions illustrated by equations (86, 89), and clinch the series with Byerly's half-humorous emphasis (Generalized Coördinates (1916), page 33). This book has the merit of helpfully discursive approach to a large subject; and though it seems tacitly limited to the vector conception, closing the matter on the range that Lagrange occupied at one bound and not gradually, proper antidote can be sought elsewhere. See Silberstein, Vectorial Mechanics (1913), page 59; while Ebert has been referred to in note 3, for his treatment in the larger spirit of energetics.

We insert now the quotation from Heaviside; it illustrates fairly the *ne plus ultra* in both respects. Notation of our text is continued. Because (E) is a homogeneous quadratic function of the velocities, Euler's theorem about homogeneous functions enables us to write

$$2E = \Sigma\left(\dot{k}\frac{\partial E}{\partial \dot{k}}\right),$$

of which the legitimate total time-derivative is

$$2\frac{dE}{dt} = \Sigma\left[\frac{d}{dt}(\dot{k})\frac{\partial E}{\partial \dot{k}} + \dot{k}\frac{d}{dt}\left(\frac{\partial E}{\partial \dot{k}}\right)\right].$$

Since (E) is "by structure" a function of velocities and coördinates only,

$$dE = \Sigma\left(\frac{\partial E}{\partial k}dk\right) + \Sigma\left(\frac{\partial E}{\partial \dot{k}}d\dot{k}\right).$$

Divide the last equation by (dt) and subtract from the second, giving

$$\frac{dE}{dt} = \Sigma\left\{\left(\frac{d}{dt}\left(\frac{\partial E}{\partial \dot{k}}\right) - \frac{\partial E}{\partial k}\right)\dot{k}\right\} = \Sigma(F_k\dot{k});$$

the last member expressing the energetic invariance of activity (see equation (298)).

It would be misleading if the text pretended to do more than give Lagrange's equations their setting of introductory connection with the other topics treated. In order to proceed safely the results here gleaned must be followed up seriously; the references given already indicate where to begin, and they can be relied upon to supplement themselves as the subject opens. Questions to be met at once are alluded to incidentally in section 136: a rationally consistent view of superfluous coördinates, including how they may drop that character and become physical; and the bearing of that quoted "interlocking" upon the significance of the term *holonomous.* That there are more vital issues awaiting analysis is suggested by Burbury (Proceedings of the Cambridge Philosophical Society, VI, page 329); by such comment as Heaviside's (Electromagnetic Theory, III, page 471) upon Abraham's successful extension of Lagrange's equations; and by the lines of inquiry to which note 32 points.

Note 31 (page 194). This development is seen to be borrowed from Thomson and Tait, pages 320–24. The few changes are adapted here and there to an even keener intent to keep the energies and momenta at the front, subordinating the investiture with mathematics. It was thought needful to drive the entering wedge before closing, for the sake of those continuations to which Maxwell's example leads. The reduction factors (l, m, n) are easily released from their trigonometrical meaning, and other geometrical implications cancelled.

Note 32 (page 200). For the justified application of equation (333), or of forms derivable mathematically from it, to all sequences of energy change, one turning-point is set by delimiting the *necessary equivalences* between the mechanical readings of (E) and (Φ) and the broader dynamical ones. This general idea is pursued by Königsberger in his papers, Über die Prinzipien der Mechanik (Sitzungsberichte der Berliner Akademie (1896), pages 899; 1173); and is entertained by Whittaker in his Analytic Dynamics (1904), Chapter X, *passim.* The stimulus to this quest seems still attached to the possibility of constructing a parallel in mechanical energy by using values connected with other energy changes. One gathers this meaning from the utterance of Larmor (Aether and Matter, page 83) and others like it.

INDEX

The Numbers refer to Pages

The Numbers refer to Pages

The Numbers refer to Pages

The Numbers refer to Pages

The Numbers refer to Pages

The Numbers refer to Pages

The numbers refer to Pages

The numbers refer to Pages

The Numbers refer to Pages

www.ingramcontent.com/pod-product-compliance
Lightning Source LLC
Chambersburg PA
CBHW021949220326
41599CB00012BA/1420
9780520376441